THE HUMAN CONTRIBUTION

The Human Contribution

Unsafe Acts, Accidents and Heroic Recoveries

JAMES REASON

Professor Emeritus, The University of Manchester, UK

LONDON AND NEW YORK

First published 2008 by Ashgate Publishing

Published 2016 by Routledge
2 Park Square, Milton Park, Abingdon, Oxon OX14 4RN
711 Third Avenue, New York, NY 10017, USA

Routledge is an imprint of the Taylor & Francis Group, an informa business

British Library Cataloguing in Publication Data
Reason, J. T.
The human contribution : unsafe acts, accidents and heroic recoveries
1. Industrial safety 2. Human behavior 3. Errors
4. Resilience (Personality trait) 5. Heroes
I. Title
363.1'1

ISBN: 978-0-7546-7402-3

Library of Congress Cataloging-in-Publication Data
Reason, J. T.
The human contribution : unsafe acts, accidents, and heroic recoveries / by James Reason.
p. cm.
Includes index.
ISBN 978-0-7546-7400-9 (hardcover) -- ISBN 978-0-7546-7402-3 (pbk) 1. System safety. 2. Errors. I. Title.

TA169.7.R43 2008
363.34--dc22

2008038859

ISBN: 978-0-7546-7400-9 (hardback)
ISBN: 978-0-7546-7402-3 (pbk)

Contents

List of Figures

List of Tables

About the Author

James Reason was Professor of Psychology at the University of Manchester from 1977–2001, from where he graduated in 1962. He obtained his PhD in 1967. From 1964–76, he was Lecturer then Reader in Psychology at the University of Leicester. He has also worked at the RAF Institute of Aviation Medicine, Farnborough and the US Naval Aerospace Medical Institute, Pensacola.

His primary research interest has been the human and organizational contributions to the breakdown of complex, well-defended systems. He has written books on absent-mindedness, human error, aviation human factors, on managing the risks of organizational accidents and, most recently, on error management in maintenance operations. He has researched and consulted in the fields of aviation, railways, nuclear power generation, maritime safety, oil exploration and production, mining, chemical process industry, road safety, banking and health care.

He received the Distinguished Foreign Colleague Award from the US Human Factors and Ergonomics Society (1995), the Flight Safety Foundation/Airbus Industrie Award for achievements in human factors and flight safety (2001), and the Roger Green Medal from the Royal Aeronautical Society for contributions to human factors as applied to aerospace (2001), and the Flight Safety Foundation/Boeing Aviation Safety Lifetime Achievement Award (2002). He is a Fellow of the British Academy, the Royal Aeronautical Society and the British Psychological Society. He received an honorary DSc from the University of Aberdeen in 2002, and was awarded a CBE in 2003 for contributions to patient safety. In 2006, he was made an honorary fellow of the Royal College of General Practitioners.

PART I
Introduction

Chapter 1

The Human Contribution: Hazard and Hero

Introduction

The purpose of this book is to explore the human contribution to both the reliability and resilience of complex well-defended systems. The predominant mode of treating this topic is to consider the human as a hazard, a system component whose unsafe acts are implicated in the majority of catastrophic breakdowns. But there is another perspective, one that has been relatively little studied in its own right, and that is the human as hero, a system element whose adaptations and compensations have brought troubled systems back from the brink of disaster on a significant number of occasions.

After studying human unsafe acts within hazardous enterprises for more than three decades I have to confess that I find the heroic recoveries of much greater interest and – in the long run, I believe – potentially more beneficial to the pursuit of improved safety in dangerous operations. Since most observations of people in high-risk systems are event-dependent; that is, they generally emerge from well-documented accident investigations, it is inevitable that we should know far more about the hazardous human than the heroic one.

But there is a stark contrast between unsafe acts and these intrepid recoveries. Errors and violations are commonplace, banal even. They are as much a part of the human condition as breathing, eating, sleeping and dying. Successful recoveries, on the other hand, are singular and remarkable events. They are the stuff of legends. But these abilities are not unattainable. A few people are born heroic, but most of us can acquire the skills necessary to give a better than an evens chance of thwarting a disaster scenario. However, it must be acknowledged that such heroism is not necessarily an enduring characteristic. Even the best people have their bad days.

The Structure of the Book

The book is organised into five parts. The present chapter introduces the subject matter. Chapter 2 offers an inside-out guide to being a mind user. We all know what it feels like to have a mind, but we don't always appreciate that 'feelings-of-knowing' can sometimes be very misleading. There are things you think you know about how your mind works, but don't. And there are things you think you don't know, but actually do. To realise your mind's full potential, you need to understand when these feelings of knowing (or not knowing) are useful and when they are deceptive. I will be discussing these mental processes from a user's perspective: from the inside looking out, not from the outside looking in – hence an inside-out view.

I'm not denying the intimate connection between the mind and the brain. It is just that this book is not about brain scans and neural wiring diagrams. While modern techniques can tell us a great deal about the brain's structure and function, they never wholly capture the moment-to-moment experiences of being a mind user, and none can track all of the subtle interactions between the conscious and automatic control processes. And that is what interests me here.

Part II is concerned with unsafe acts: errors and violations, and how they are perceived by those upon whom they impact. Unsafe behaviour may be less fascinating than heroism, but it is no less important.

Chapter 3 focuses on the nature and varieties of human error. In order to limit the damaging occurrence of errors and improve their chances of detection and recovery, we need to understand something of their cognitive origins and the circumstances likely to promote them. This understanding can translate into 'error wisdom' at the sharp end – what has been termed 'individual mindfulness', to be discussed in Part V.

Chapter 4 deals with rule-related behaviour. I begin by considering the various types of violation, and then discuss the social, emotional and systemic factors that lead people to choose not to comply with rules, regulations and safe operating procedures. However, such acts of non-compliance are not universally bad. They can have beneficial as well as unwanted consequences. This becomes evident when we look in detail at the 12 varieties of rule-related behaviour.

Chapter 5 examines a number of different perceptions of human unsafe acts, of which the two most dominant are the person and the system models. Each has its own theory of how these unsafe acts arise, and how they might be remedied and managed. The person model, asserting that errors originate within the minds of the people concerned, is intuitively appealing and still holds sway in many domains. However, over recent years, the system model has gained increasing ascendancy. This argues that the people on the frontline are not so much the initiators of bad events as the inheritors of long-term system failings. My thesis is that the extremes of both views have their shortcomings. We need to strike a balance between the two.

Part III deals with accidents and their investigation. One fact that lends strong support to the system approach is that similar situations keep provoking the same kinds of unsafe acts in different people. These recurrences, discussed in Chapter 6, indicate that a substantial part of the problem is rooted in error-provoking situations rather than in error-prone people. A primary function of error and incident reporting systems is to identify these 'error traps'. Eliminating them becomes a priority task for error management programmes.

Complex hazardous systems are subject to two kinds of bad event: individual accidents, resulting in limited injury or damage, and organisational accidents that occur relatively infrequently but whose consequences can be both devastating and far-reaching. One of the features that discriminates between these two kinds of adverse event is the degree of protection available against the foreseeable hazards. Whereas individual accidents usually result from the failure of very limited safeguards (or their absence), organisational accidents involve a concatenation of breakdowns among many barriers, safeguards, and controls. It is this combined failure of the many and varied 'defences-in-depth' that characterises the organisational accident and it is this type of event that will be the main concern of this book.

Chapter 7 focuses on two pioneering accident investigations that fundamentally changed the way the human contribution to bad events is regarded. In particular, they spelled out how unsafe acts and latent organisational conditions (resident pathogens) interact to breach the multi-layered system defences. It also traces how the emphases of investigations have shifted from technical and human failures at the sharp end to examining the effects of 'upstream' factors such as organisational processes, safety culture, regulation and even the economic and political climate. It is suggested that perhaps the pendulum has swung too far towards identifying causal factors that are remote in time and place from the local events. This chapter also looks at some of the problems facing accident investigators, and others who seek to make sense of the past. One such problem is the failure to differentiate between conditions and causes, thus falling foul of the counterfactual fallacy

For these and related reasons, it is argued that continual tensions between production and protection lead to resident pathogens being seeded into the system, and this is true for all systems. But such organisational shortcomings are conditions rather than causes. Although they contribute to defensive failures, they are not in themselves the direct causes of accidents. The immediate triggers for such bad events are local circumstances: human and technical failures that add the final ingredients to an accident-in-waiting that may have been lurking in the system for many years. All systems, like human bodies, have resident pathogens. They are universals. It is usually only the proximal factors, immediate in both time and space to the accident, that distinguish between a system suffering a catastrophic breakdown and those in the same sphere of operations that do not.

Up to this point, the book deals mainly with the human as a hazard. In Part IV, we look at the other side of the coin: the human as hero. Eleven stories of heroic recovery are told. They are grouped into four chapters:

- Chapter 8 (training, discipline and leadership) examines two military case studies: the retreat of Wellington's Light Brigade on the Portuguese–Spanish border in 1811; and the retreat of the US 1st Marine Division from the Chosin Reservoir in 1950.

- Chapter 9 (sheer unadulterated professionalism) deals with Captain Rostron and the rescue of the *Titanic* survivors in 1912; the recovery of *Apollo 13* in 1970; the Boeing-747 Jakarta incident; the recovery of the BAC 1-11 in 1990; and surgical excellence as directly observed in 1995–96.
- Chapter 10 (luck and skill) looks at the near-miraculous escapes by the 'Gimli Glider' on the edge of Lake Winnipeg in 1983 and United 232 at Sioux City in 1989.
- Chapter 11 (inspired improvisations) discusses General Gallieni and the 'miracle on the Marne' in 1914; and the saving of Jay Prochnow lost in the South Pacific by Captain Gordon Vette in 1978.
- What, if anything, did these heroes have in common? Chapter 12 seeks to identify the principal ingredients of heroic recovery.

Part V (Achieving Resilience) has two chapters. Chapter 13 elaborates on Karl Weick's notion of 'mindfulness'. In its broadest sense this involves intelligent wariness, a respect for the hazards, and being prepared for things to go wrong. Mindfulness can function both at the level of the frontline operators and throughout the organisation as a whole. The former we term 'individual mindfulness' and the latter 'collective mindfulness'. Both are necessary to achieve enhanced systemic resilience. We can't eliminate human and technical failures. And no system can remain untouched by external economic and political forces. But we can hope to improve its chances of surviving these potentially damaging disruptions in its operational fortunes.

The last chapter deals with the search for safety, the broadest part of the book's spectrum. Two models of safety are described: the safety space model and the knotted rubber band model. The former operates at the cultural and organisational levels; the latter has a more tactical focus and deals with keeping some continuous frontline process within safe boundaries. Together, they have implications for re-engineering an existing culture to improve safety and resilience. Or, to put it another way, this concluding chapter is concerned with the practical measures necessary to achieve states of both individual and collective mindfulness.

About the Book

I should end this introduction by saying something about the readership and style of the book. Perhaps I should begin by saying what it is not. It is not a scientific book, even though it touches upon scientific and technological issues. It is written in the first person. That means that these are my personal views – and prejudices. It is written for real people, even though academics and students may find parts of it of interest. But it does not require any prior knowledge of academic psychology, although these issues are touched upon in the first chapters. Nor is it a 'how-to-do' book. If there is any way of describing the content, I would say it was about the philosophy of managing complex hazardous systems. Philosophy is a daunting word, but in this book it simply means a way of thinking about the issues. In short, it is a way of confronting the problems of conducting a hazardous operation so that you keep your risks as low as reasonably practicable and still stay in business. It is this latter injunction that, for me, is the most important one.

Chapter 2

A Mind User's Guide

After using a mind for seventy years, I realise that I know very little about it, and the older I get the more this conviction grows. It is true that after nearly forty years of researching and teaching psychology, I do have an inkling of what I do and don't know. It is also the case that I have some understanding of what I think I know, but really don't. And, if I ask myself the right questions, I can occasionally come up with things that I didn't think I knew, but actually did. For all that, though, much of my mental functioning remains secretive, seemingly out of reach and full of surprises.

But – you might be thinking – Sigmund Freud told us that over a hundred years ago.[1] So what is new? Quite a lot, as I will hope to show later. Freud is closely linked with the idea of an unconscious mind, but he did not invent the term, nor is his rather narrow view of the unconscious widely accepted by contemporary psychologists. I do not reject the idea of an unconscious mind – indeed, its existence is the main reason for including this chapter – but I do dispute the strict Freudian interpretation of its role in our mental lives.

My purpose here is to make you, the everyday mind user, more familiar with the mysteries of your own mental life, and – once in a while – to tell you something that you did not already know. This chapter is also intended to act as an introduction to the discussion of errors and violations in the next two chapters.

Tip-of-the-Tongue State

Let's start with a commonplace experience. There's nothing quite like the 'tip-of-the-tongue' (TOT) state to expose the subtleties

1 Freud, S. (1914) *Psychopathology of Everyday Life*. London: Ernest Benn. (Originally published in 1901).

of knowing and not knowing the things that go on in your own mind. A TOT state begins with an attempt to retrieve from memory something that you are sure you know, but then the search fails to yield an immediate, felt-to-be-correct response. Instead, it produces a name or a word that you recognise as being 'warm', but you also know that this is not the sought-for item.

When you persist with the search, the same wrong item keeps coming to mind in an irritatingly obtrusive fashion. What makes the whole experience so frustrating is that this recurrent blocker is felt to be very close to the target item. We appreciate that it might have similar properties, like sound, structure or meaning, but yet we are certain it's wrong. How do we know these things when we can't directly access the right word or item? Some part of the mind knows, but it's not the conscious part.

Here is an actual example. I was searching for the name of D.W. Griffith's silent film *Intolerance*. Every time I made an active memory search I came up with 'intemperance'. I knew this was wrong, but it felt very close. I had some conscious information about the sought-for word – I knew it was a single-word title, I had a rough idea of the number of syllables, and I had a strong feeling that it began with 'I'. But what I did not know at the outset was that the word had an '*-erance*' ending, though it was clear that some part of my mind knew this by the repeated retrieval of 'intemperance'.

The search experience was rather like standing on a small stage – equating to my conscious awareness – with large wings on either side. I would go to one of the wings and call out the search cues: it's the title of a classic film; it begins with 'I'; it has many syllables; it covered a wide historical sweep of man's inhumanity to man; and so on. Then 'intemperance' would be repeatedly thrown on to my conscious stage, and a voice from the other wing would call 'no, but you're very close'. Somewhere off-stage was the correct template for the sought-for word, but I couldn't access it directly. I only recognised it when the search finally produced the right answer, *Intolerance*, which I immediately knew was correct.

The Conscious and the Automatic Modes of Control

This TOT experience shows that one of the main problems with the human mind is that the user is only in direct conscious contact with a fraction of the whole. The conscious part seems to be located somewhere between the ears and behind the eyes. At any one moment in this very limited space the larger part of our current waking thoughts and feelings are experienced, our sense data interpreted and our present actions planned, initiated and monitored. It is also this tiny space that feels at that instant most closely identified with our innermost selves – our personal beliefs, attitudes, values, memories, likes and dislikes, loves and hates, and the other passing clutter and baggage that goes to make up one's mental life. But we are only aware of a very restricted amount at any one time. The ideas, feelings, images and sensations seem to flow like a stream past a blinkered observer standing on the bank. We can't see far upstream or downstream, but we can take in between one to two seconds worth of what goes past. This is what comprises the conscious workspace, the experiential here and now.

Beyond this present awareness lies a vast and only partially accessible knowledge base. Some of the information contained there is in the form of previous life events (episodic memory), though this becomes very patchy for the time before we were five years old, maybe even later. Other parts of it are used to make sense of the world (semantic memory). And yet other knowledge structures (called schemas) control our routine perceptions, thoughts, words and actions.

We have a rough idea of the contents of this long-term knowledge base – not all of them, of course, but enough to be aware of the general headings. But what we don't know is how stored items are called to mind. Such retrievals can be so accurate and so immediate as to convince us – incorrectly, as I hope the TOT example has shown – that we have direct voluntary access to all parts of the store. The reality is that while we are conscious of the products of such retrievals – the words, feelings, images, thoughts and actions – we have little or no awareness of the processes that seek them out and call them to mind. Understanding this is very important since most of our mental lives involve a continuous interaction between the conscious workspace and the long-term memory store. Sometimes we deliberately call items to mind, but at other times they simply pop up unbidden.

The continuous interplay between the conscious workspace and long-term memory or knowledge base will occupy us for the remainder of this chapter. These two co-existing, and sometimes competing controllers of mental life have markedly contrasting properties. These properties are summarised in Table 2.1.

Though these two mental components work in harmony for much of the time, they can also compete for command of the body's output mechanisms, both in the observable physical world, through unintended words and actions, and in the conscious workspace, into which items may be delivered without conscious intent. This is hardly surprising given their radically differing properties and the power of familiar environments to evoke habitual responses.

Table 2.1 Comparing the properties of the conscious workspace and the long-term knowledge base

Conscious Workspace	Long-term Knowledge Base
Accessible to consciousness. Closely linked with attention and working memory.	While the products (actions, thoughts, images, etc.) are available to consciousness, the underlying processes are largely outside its reach.
Selective and resource-limited.	Apparently unlimited in both the amount of stored information and the length of time for which it is retained.
Slow, laborious and serial (one thing after another).	Fast, effortless and parallel (many things at once).
Intermittently analytical. Sets intentions and plans and can monitor their progress at the various choice points. But this often fails.	Automatic in operation.
Computationally powerful. Accepts inputs from nearly all senses. Vision dominates.	Behaviour governed by stored specialised knowledge structures (schemas) that respond only to related sensory inputs and do their own thing.
Accesses long-term memory by generating 'calling conditions' or retrieval cues.	Two basic retrieval processes: similarity-matching (like with like), and frequency-gambling (resolving possible conflicts in favour of the most frequent, recent or emotionally charged items).

Three Levels of Performance

The extent to which our current actions are governed either directly by conscious attention or more remotely by pre-programmed habit patterns gives rise to three levels of performance: knowledge-based, rule-based and skill-based. Their contrasting characteristics are summarised in Figure 2.1.

All human performance – with the exception of what comes 'hard-wired' at birth – begins at the knowledge-based level in which our actions are governed online by the slow, limited, and laborious application of conscious attention. This level relies very heavily upon conscious images or words to guide our actions, either in the form of inner speech or through the instructions of others. While this type of control is flexible and computationally powerful, it is also highly effortful, enormously tiring, extremely restricted in scope and very error prone – and we don't like it very much.

Although we all know what attention feels like, its precise function in mental life is not at all obvious. An optimum amount of attention is necessary for successful performance in all spheres of activity; but too little or too much can be highly disruptive. The consequences of inattention are clear enough, but if you need an example of over-attention, try using a keyboard while thinking about what the index finger of your right hand is doing

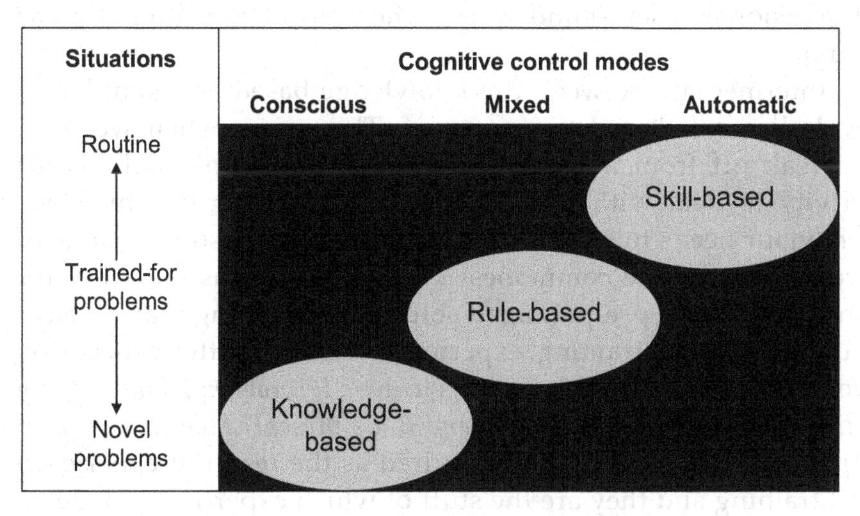

Figure 2.1 Three levels of performance control

– the greater your typing skills, the more likely it is that this will cause problems. As in many other areas of psychology, we can learn a good deal about the function of attention by observing its occasional failures.

At the other end of the spectrum there is skill-based performance. By dint of practice, self-discipline and the reshaping of our perceptions, we can gradually acquire the rudiments of a skill – that is, the ability to mix conscious goal-setting and guidance with the largely automatic control of our individual actions. Instead of agonising over each separate movement or word, we are able to run them off in pre-packaged sequences. This is the stuff of which habits are made, and it is the essence of skill-based performance. William James wrote: 'Habit diminishes the conscious attention with which our acts are performed.'[2]

This gradual automatisation of performance is universal; it occurs in all areas of mental performance. Even our social interactions become more and more automatic with time and experience. Without this ability to delegate control to non-conscious habit sequences, or motor programmes, we would consume all our limited attentional resources in dealing with the present moment and have nothing left to review the past or plan for the future. Such perpetual 'present-mindedness' would be insupportable; we could spend all day trying to tie our shoe-laces. But nothing comes free. Automatisation carries the penalty of occasional absent-mindedness when our actions do not go as planned.

Intermediate between the knowledge-based and skill-based levels lies rule-based performance. This arises when we need to break off from a sequence of largely habitual (skill-based) activity to deal with some kind of problem, or in which our behaviour needs to be modified to accommodate some change of circumstances. The commonest kinds of problems are those for which we have a pre-packaged solution, something that we have acquired through training, experience or some written procedure. We can express these solutions as 'rules': *If [problem X] then [apply solution Y]*, or *If [indications A and B are present] then [it is a type-C problem]*. These 'rules' are acquired as the result of experience and training and they are the stuff of which expertise is made.

2 James, W. (1889) *Principles of Psychology*. New York: Holt (p. 114).

However, as we shall see in the next chapter, this level of performance is associated with a variety of errors. We can misapply a normally good rule (i.e., one that usually works in this particular situation) because we have not noticed the contra-indications; we can apply a bad rule; or we can fail to apply a normally good rule – a mistaken violation.

When we run out of pre-programmed problem solutions, as in some novel or unexpected situation, we are forced to resort to working out a solution 'on the hoof' using the slow, effortful, but computationally powerful conscious control mode. This is a highly error prone level of performance and is subject to a range of systematic biases. These knowledge-based mistakes will be considered in the next chapter.

Interacting with the Long-term Knowledge Base

The secretive properties of long-term memory and, in particular, the processes by which stored items are called to mind lie at the heart of the mind-user's misunderstandings of his or her mental function. There appear to be a number of different mechanisms involved in memory retrieval and two of them – similarity-matching and frequency-gambling – are automatic, unconscious and continuously operative. Since we cannot introspect upon these processes directly, we can only guess at their nature by observing the recurrent patterns shown by our errors, and by which of these processes, similarity or frequency, dominates in different kinds of memory search.

When the initial search cues are detailed or highly specific, matching these calling conditions to the characteristics of stored memory schemas on a like-to-like basis is the primary retrieval process. However, when the search cues match several stored knowledge structures, the mind gambles that the most frequently used knowledge item in that particular context will be the one that is required. Two examples will make the point clearer. If we were asked what is it that barks, has four legs, wags its tail, cocks it leg at lampposts and is regarded as man's best friend, most of us would quickly retrieve the knowledge item that matches all of these characteristics uniquely – namely a dog. Here the retrieval is based almost entirely on similarity matching and converges upon a specific stored item. The process is so rapid that we feel we have reached out and retrieved the item in a conscious and deliberate fashion.

However, if we were asked to generate exemplars of the category 'four-legged animal' in no particular order, it is highly likely that, on average across a group of people, the first items coming to mind would be dog, cat, horse and cow. Once again, the item 'dog' is called to mind, but the search process was not similarity-matching. In this case, the order of retrieval is dominated by the familiarity of the animal. Familiarity is a function of frequency of encounter, so, in this divergent memory search, frequency-gambling is the primary search process.

Memory searches are strongly influenced by 'feelings of knowing'. We do not strive to retrieve things that we know we don't know. But we doggedly continue searching for something that we are sure we know, even though, as discussed earlier, we keep coming up with the wrong item. So, from the mind-user's point of view, these feelings about the contents of memory are of considerable value. They are not always right, of course, but they are correct often enough for us to treat them as handy guides to whether or not we should invest mental effort in a memory search. There are also things we don't always realise we know, but actually do – if only in a very approximate way.

A good example of this is frequency of encounter. The research evidence suggests that people automatically log how often they come across any reference to a particular person or topic. This logging process does not involve an actual numerical count; it is best captured by asking people to rate how often they have encountered some person or thing on a graded 0–6 scale ranging from 'never' (0) to 'nearly all the time' (6). By the same means, we can also obtain a fairly good assessment of co-occurrences. That is, we can gauge in very general terms how often X has occurred with Y with a moderate degree of accuracy. Understanding and exploiting these feelings of knowing is very useful, though not always appreciated by the mind user.

The human mind is exceptionally good at simplifying complex information-handling tasks. It does this by relying as far as possible upon the automatic mode of control and using intuitive 'rules of thumb' or heuristics. These are unconscious strategies that work well most of the time, but they can be over-utilised and produce predictable forms of error. We have already mentioned two of these heuristics: matching like-to-like (similarity-matching) and resolving any competition for limited conscious access by favouring the most frequently encountered candidate (frequency-gambling).

Under conditions of stress or danger, we are inclined to fall back on well-tried solutions or past plans rather than ones that might be more appropriate for the current circumstances. The moral here is: be wary of the commonplace and familiar response for that particular situation. It may indeed be appropriate, but it needs to be considered carefully because this is exactly the kind of response that the automatic heuristics are designed to produce. The automatic mode of control is shaped more by the past than the present. There is always the danger of a 'strong-but-wrong' reaction – a very common type of human error as we shall discover in the next chapter.

Intentions and the Retrieval Cycle

I have argued that consciousness has no direct access to the long-term knowledge base, only to its products (ideas, actions, images, words and the like). Its sole means of directing knowledge retrieval is through the manipulation of calling conditions or the search cues that arise from either the outside world or from conscious 'mindwork'. The searches themselves are carried out unconsciously by two automatic processes: similarity-matching and frequency-gambling. All that the conscious workspace can do in this regard is to deliver the initial calling conditions, assess whether the search products are appropriate and, if not, reinstate the search with revised calling conditions. If that is the case, what gives the mind its intentional character – one of the defining features of adult mental life? In short, how does it instigate and direct goal-directed behaviour?

As with many other questions relating to mind use, William James provided an answer: 'The essential achievement of the will ... is to attend to a difficult object and hold it fast before the mind. The so doing is the fiat; and it is a mere physiological incident that when the object is thus attended to, immediate motor consequences should ensue.'[3]

This statement maps directly onto the properties of the retrieval cycle described earlier. The 'holding-fast-before-the mind' results in a sustained series of related calling conditions. The consistency of these cues creates a high level of focused activation within a limited number of knowledge structures or schemas. This will automatically release their products to consciousness and the effector organs (that is, our limbs, speech centres, thought processes, and the like).

But the persistence of these same-element cues has to be maintained in the face of continual pressure from other claimants to the very limited conscious workspace. This attentional effort can only be sustained for short periods, particularly when the goal is unappealing (e.g., getting out of bed on a cold morning) or when the objects being attended to are boring (e.g., listening to a dull lecture). Under these circumstances, as James put it, 'our minds tend to wander (and) we have to bring back our attention every now and then by using distinct pulses of effort, which revivify the topic for a moment, the mind then running on for a certain number of seconds or minutes with spontaneous interest, until again some intercurrent idea (or image) captures it and takes it off.'[4]

Concurrent Processing

The human mind has an extraordinary ability to store the recurrences of the world in long-term memory as schemas (knowledge packages), and then to bring their products into play whenever they correspond to the current contextual calling conditions.[5] We are furious pattern matchers; it's what our minds do best. And we do this without recourse to the computationally powerful but very limited operations of the conscious workspace.

3 James, W. (1890), p. 561.

4 James, W. (1908) *Talks to Teachers on Psychology and to Students on Some of Life's Ideals*. London: Longmans, Green & Co. (p. 101).

5 See also Reason, J. (1990) *Human Error*. New York: Cambridge University Press.

This does not mean that conscious operations play no part in the retrieval cycle, simply that long-term memory is capable of spitting out its products (thoughts, images, words, etc.) without the necessity of higher-level direction. Inference and other conscious processes act upon the products of long-term memory retrieval, and this mindwork then creates a new series of calling conditions. We can best illustrate this concurrent processing by considering the steps involved in trying to solve a problem. Before doing this, however, it is necessary to remind ourselves of some important differences between the conscious workspace and the schemas in long-term memory in the way they accept information from the outside world.

Consciousness keeps open house to all kinds of data, either externally or internally generated. But each knowledge structure within long-term memory is tuned a highly specific set of triggering conditions and is largely oblivious to all that falls outside this very restricted range of inputs. Thus, a cat schema is only interested in strictly feline issues; a mother schema responds only to maternal issues. Each schema automatically scans the world for those inputs that match its own very parochial concerns.

Every problem we encounter broadcasts an array of retrieval cues. These are detected both by the relevant schemas within the knowledge base and by the conscious workspace. Almost immediately, a candidate solution or hunch is delivered to the conscious workspace through the automatic intervention of similarity-matching and frequency-gambling. On first sight, this seems to fly in the face of received wisdom in which answers to problems are supposed to follow analysis and inference. In reality, however, it is usually the other way round. A possible answer springs to mind automatically (or intuitively), and then we do the more laborious mindwork to establish whether or not it is correct.

Let's suppose that on this occasion we reject the solution. This conscious inferential activity generates a revised set of retrieval cues that are subjected to further automatic processing by the similarity-matching and frequency-gambling heuristics and another possible solution pops into mind. Again, we evaluate it using conscious processing. If we accept the solution as being appropriate, it is acted upon. If it is rejected a further set of automatic processes are set in train – and so the retrieval cycle proceeds until a satisfactory answer is found.

I mentioned earlier that stress and strong emotion can lead to 'strong-but-wrong' responses – that is, behaviour that is a frequent and generally useful feature of a particular situation, but which, on some occasions, is inappropriate or incorrect. The retrieval cycle just described provides an explanation for this, perhaps the commonest error form. The conscious workspace is very limited in its capacity to handle information and can easily be overwhelmed by the press of unrelated events, causing the cycle to be halted prematurely.

Military trainers have long understood that battlefields are not conducive to slow and effortful conscious thought. Essential problem-solving 'rules' – like the actions required for clearing a jammed weapon – are drummed into soldiers' heads so that the behaviour is carried out in an almost reflexive fashion when the need arises.

The Relationship Between Memory and Attention: The Blob-and-the-Board Model

This book includes a number of models based on easily pictured and remembered images: the *blob-and-the board,* to be discussed below; the *Swiss cheese model* relating to organisational accidents (see Chapter 5); the *three-bucket model* concerning 'error wisdom' (Chapter 13); and the *knotted rubber band model* that shows what is necessary to keep a system within safe and reliable limits (Chapter 14). Models of this kind are not complete, or even adequate explanations, but they serve to capture complex interactions in a simple, memorable and visual way. Experience tells me that this is what people take away from my presentations and writings. Metaphors are powerful persuaders.

Figure 2.2 shows an amorphous blob located over a limited number of squares on something that looks like a chess board. The blob represents the limited attentional resource. It is constantly skidding around over the board. Its presence over a particular square on the board is necessary to realise the full potential of the underlying knowledge structure or schema. The squares represent individual knowledge packages within long-term memory. Each

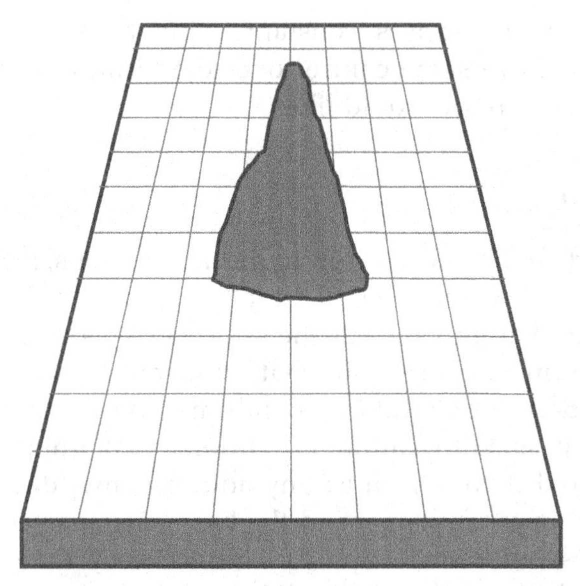

Figure 2.2 The 'blob-and-the-board' model

one relates to a particular aspect of our stored memories: habits, routine actions, thoughts, words, social skills, perceptions, semantic and episodic (our own experiences) schemas, and so on. The longer and the more concentrated its presence, the greater will be the likelihood that matters relating to the underlying schema will be called to mind or put into action, unbidden or not.

If we are preoccupied or depressed, the blob will take on a narrow peaky aspect, locating most of its finite resources over a specific schema. This means that we will be subject to repetitive ruminations: the same thoughts will keep coming to mind. Or, when we are distracted, it could mean that this schema will prompt unwanted and unintended actions. This is what happens to absent-minded professors when they are continuously abstracted by sustained pondering on some very specific issue.

At the other extreme, the blob spreads thinly over several schemas like a badly laid rug. In this case, we cannot concentrate on any one thing. We are not able to realise the full potential of the underlying knowledge structures. Our minds flit from one thing to another as the blob skids around the board, our actions are not focused on any particular goal, and we cannot easily follow a consistent plan of action.

Mental life comprises constant shifts between these two extremes. We have some control over the location of the 'blob', but not as much as we would like.

Summing Up

Leaving aside the Freudian view of the unconscious, most of what we think we know but generally don't know about the workings of our minds has to do with the way we retrieve information from long-term memory. And that, after all, is how we spend the greater part of our waking 'mindtime'. When we supply the knowledge base with calling conditions that uniquely match a single stored item – such as 'my address', my 'date of birth', 'my mother's maiden name' and the like – the answers arrive in consciousness with such rapidity and accuracy that we can be excused for believing we have direct control over the retrieval process. And so we do – but only up to a point.

That this is not the general rule is shown by experiences like the TOT state and the fact that things are continually popping into consciousness without any direct bidding. Between consciously initiating memory retrieval and receiving the result lie the automatic and unconscious search processes of similarity-matching and frequency-gambling.

Since these processes are unconscious and not immediately evident to brain scans and imaging techniques, you might reasonably wonder how we know about their existence. They are, like the knowledge structures themselves, theoretical constructs that are necessary to explain memory's oddities (like the TOT state) and the recurrent nature of our 'strong-but-wrong' errors.

They can also be guessed at by the results of category generation, a technique whereby we ask people to generate exemplars of categories like 'trees', 'gemstones', 'four-legged animals' and 'American presidents'. Another method is by asking people 'Trivial Pursuits' questions of the 'who said?' kind. Here are two examples:

1. When asked who said 'The lamps are going out all over Europe; and we shall not see them lit again in our lifetime' most people (over 90 per cent in British audiences) respond with Winston

Churchill. It was actually said by Sir Edward Grey, the Foreign Secretary in 1914. The Churchill answer is wrong, but it follows a predictable and sensible pathway. The calling conditions of the quotation suggest that it was said by an English-speaking statesman on the verge of something cataclysmic, like the Second World War. Churchill was, for most people, the most frequently encountered politician of that time – and he was certainly the gabbiest. And hardly anyone remembers Sir Edward Grey. If we can predict an error with a high degree of confidence, we begin to understand the hidden processes that gave rise to it.

2. In the 1980s, when I was researching this issue, I would ask my lecture audiences who said 'the buck stops here' (the words Truman had on a sign on his desk). The calling conditions suggest it was said by an American because of 'buck'. Who are the most famous Americans? Presidents. At that time Reagan was in the White House and he was the fortieth US president. Whereas Americans could, on average, recall the names of around 32–36 presidents, British psychology students could hardly do better that eight to ten. But in both cases, the frequency gradients were much the same. The frequency-of-encounter data followed a roughly U-shaped pattern with a peak in the middle. The most recent presidents are the most salient, dwindling away through the early decades of the twentieth century with minor peaks for the two Roosevelts, and then a major peak for Abraham Lincoln. And, of course, everyone remembered George Washington and maybe his immediate successors. The fact that we are not entirely slaves to frequency-gambling is indicated by the fact that relatively few respondents answered with Reagan, incumbent at that time and the most often encountered one. When pressed on this, they said 'The quotation sounds a bit hicky, and Reagan isn't a hick'. The most popular response was Nixon, a little further down the frequency gradient. When asked why, the students said something like 'Well, Nixon was a crook and the buck [which they wrongly took to be a dollar] ended up in his pocket.' The point about this lengthy explanation is that it bears out the retrieval cycle discussed earlier. Reagan probably sprang to mind first, but was rejected by not very accurate mindwork in favour of Nixon.

Mention of frequency-of-encounter brings us directly to the other side of the knowing coin: the things you think you don't know, but actually do – even though this depends upon being asked the right kind of question. There is strong evidence to indicate that we log, in a very approximate way, the number of times we encounter something, either in thought or deed. This logging process is automatic and intuitive.

If I asked you how often you had encountered any mention of, say, Indira Ghandi (any such public figure would serve as well), you'd probably say that you had no idea. But, if I asked you to make a rating on a scale of zero (not at all) to six (nearly all the time), you would hum and ha a bit, but if I pressed you to make a rapid estimate, you may well respond with something like one or two – meaning 'it's a name I've heard of, but not very often'.

When we compare these 0–6 frequency ratings with actual encounters, we find a very close agreement. For example, one of my research students showed a string of videos arising from the 1960s TV version of the *Forsyte Saga* to a group of young people who had neither seen this production nor read the book. At the end, they were given a list of the main characters and asked to make 0–6 ratings on how often they had appeared on screen. The correlation between their mean ratings and the actual times on screen was around 0.9, an almost perfect correspondence.

People have likened long-term memory to a library. This is a poor analogy. It is actually more like heaps of files strewn over a garage or warehouse floor. Let us assume that each pile relates to a particular item in our personal world. Every time we encounter the item in question, we add another file to the heap. If asked the right question (make a quick intuitive estimate), we can gauge the height of a particular pile. We can even make rapid judgements about items that are not there. It doesn't take people long, for example, to say that they have never encountered anyone called 'Ebenezer Blodderskin' (or some other unlikely name). A computer, on the other hand, would need to make a serial search. But we can establish an absence in a blinking of the mind's eye.

However, if memory has any resemblance to a library it is to some huge public library like the Manchester Central Reference Library or the British Library, where we have no direct access to the stacks. To retrieve something, we have to write out a request slip and hand it to an attendant who then fetches it for us. The attendant, in this case, stands proxy for the secretive search processes of similarity-matching and frequency-gambling.

This analogy captures something of the search process, but even this is wide of the mark. In a library, we try to provide as much information as possible about the book we want: the author, year of publication, title, publisher, and the like. We can use our memories in this organised and directed fashion, but mostly we don't even know that we've initiated a search, or when we do, the search characteristics are vague, incomplete or even incorrect – as the TOT state demonstrates.

Finally, there is the issue of guessing. When people are asked a general knowledge question and don't know the answer, but are pressed to respond, they say they were guessing. But guessing implies a random process, and there's nothing random about frequency-gambling. However, they are generally unaware of the power of frequency in shaping their answers, even though – if asked the right questions – they can give a fairly accurate mapping of the frequencies of encounter associated with a particular category of knowledge. So this lies midway between not knowing what you think you know and actually knowing what you think you don't know. A tricky chap the mind, but its hidden aspects can be made more apparent by the nature of our errors, as I will show in the next chapter.

PART II
Unsafe Acts

Chapter 3

The Nature and Varieties of Human Error

Defining and Classifying Error

Although there is no one universally agreed definition of error, most people accept that it involves some kind of deviation. Such deviations could be from the upright (trip or stumble); from the current intention (slip or lapse); from an appropriate route towards some goal (mistake); or, in some circles, it could even involve straying from the path of righteousness (sin).

Just as there are several possible definitions, so there are also many ways in which errors may be classified. Different taxonomies serve different purposes. These depend upon which of the four basic elements of an error – the intention, the action, the outcome and the context – is of greatest interest or has the most practical utility.

A Classification Based on Intention

As a cognitive psychologist studying the mental processes that give rise to error, I initially favoured a classification based upon intention. With an intentional taxonomy, different types of error can be distinguished as follows:

- Was there a prior intention to act? If not, then this was an involuntary action or automatism rather than an error.
- If intended, did the actions go as planned? If not, we are talking about absent-minded slips and lapses – failures in either the execution or the storage stages of an action sequence.
- If the actions were as intended, did they achieve their desired outcome? If not, then this is likely to be a mistake involving some failure in the planning process. The plan of action did not attain the goal because it was inadequate in some respect. This could

be in regard to the nature of the actions, or in the assessment of the situation, or both. Whereas slips and lapses occur at the level of execution, mistakes arise during the much more complex process of making plans. As indicated in the previous chapter, mistakes can be either rule-based or knowledge-based. We will consider this distinction further at a later point.

- If successful (that is, the intended outcome was achieved), did the actions involve some deliberate deviation from standard operating procedures? If so, we are dealing with some kind of violation. We will discuss these non-compliances in the next chapter.

Error Detection

Error types differ markedly in their ease of detection and recovery. This depends upon the criterion or standard against which the performance is judged. Trips and stumbles are detected and recovered automatically because our position and motion senses are 'hard-wired' to detect and compensate for deviations from the gravitational vertical.

Slips, involving 'actions-not-as-planned', are relatively easy to detect because we generally know what our current intentions are, or the circumstances make the deviation obvious. We soon recognise, for example, that getting into the bath with our socks on, or saying 'thank you' to a coffee machine, was not a part of the plan. Some memory lapses are also readily discovered. Finding yourself unable to retrieve the name of person you know well leads to instant embarrassment. But realising that you have forgotten to do something is more variable in its detection time depending on what is that was omitted. You may have to wait months to discover that you had forgotten a dental appointment, while shopping without bringing your purse or wallet is discovered the moment you stand at the checkout.

Mistakes, on the other hand, are often difficult – and sometimes impossible – to discover because we do not necessarily know the ideal pathway towards an intended goal. Not achieving the desired outcome does not, of itself, signal a mistake. A plan has two elements: a process (knowing, judging, deciding and the like) and an outcome (desired or not desired, good or bad). A process that most people would judge as sound can lead to an undesired result because of circumstances beyond the control of the planner. Conversely, an inadequate plan can yield a positive outcome because of the intervention of unforeseeable good fortune. Chance, lucky or unlucky, is even-handed; it does not distinguish between the deserving and undeserving.

Error Classifications Based on Action

On some occasions, it is more useful to consider the nature of the actions involved in the error rather than their cognitive antecedents. Action-based taxonomies can be generic, appropriate for all activities or task-specific. One possible generic classification is set out below:

- *Omissions*: a necessary or planned-for step is not done at the intended time.
- *Intrusions*: the appearance of unwanted or unintended actions, often recognisable as being part of some other activity.
- *Repetitions*: unnecessarily repeating actions that have already been performed.
- *Wrong objects*: the right actions are carried out, but in relation to the wrong objects.
- *Misorderings:* the right actions are carried out, but in the wrong sequence.
- *Mistimings*: the right actions are carried out but at the wrong time.
- *Blends*: the unintended merging of two action sequences meant to serve different goals.

In a task-specific classification, the categories are expressed in technical terms relating to the domain in question. Surgical errors, for example, can be classified according to what went awry: wrong blood vessel, wrong nerve or duct, wrong organ, wrong site or side, tissue tearing, bad knots, unchecked bleeding, and the like.

One of the main advantages of action-based classifications is that there is generally a high degree of consistency among the classifiers. In other words, there is a fair degree of inter-judge agreement as to which category a particular error belongs.

A major drawback of such schemes is that they give little or no clue as to the underlying processes. In order to carry out necessary actions at the right time, in the right order and in the right place, some or all of the following cognitive stages need to be completed correctly:

- *Plan formulation*: the intention to carry out an action must be formulated and then scheduled to be carried out at the appropriate place and time. These actions must, of necessity, be seen as contributing to the achievement of a particular goal.
- *Intention storage*: although some actions may be carried out immediately, it is more often the case that intentions to act are stored in prospective memory, and then reactivated at the appropriate time and place.
- *Execution*: the actions must be initiated and performed as planned.
- *Monitoring*: periodic attentional checks should be made to ensure that the action sequence is proceeding as intended.

An omission can have its origins at any one of these stages. The need for the action can be disregarded during plan formulation; the intention to act can be lost from storage – failures of prospective memory are very commonplace; the act can left out of the intended behavioural sequence during its execution; and its absence can escape notice during an attentional check on progress. The multiplicity of these various possible breakdowns provides strong grounds for predicting that omissions are likely to be the single most frequent error type – as indeed is the case.

Error Types Based on Contextual Factors

The situation in which an error occurs is at least as important as its psychological antecedents (if not more so) in triggering its occurrence and shaping its form. We cannot easily change human cognition, but we can create contexts in which errors are less likely and, when they do occur, increase their likelihood of detection and correction. As indicated in the previous chapter, situations can be more or less error-provoking. Here is a list of some contextual error types:

- *Anticipations and perseverations*: errors can be clearly shaped by what is coming up and what occurred previously in an action sequence. Actors and newsreaders are especially prone to these kinds of errors. Actors, for example, can be triggered into uttering lines belonging to a later act. This is an anticipation error due to similarities in content, sound or circumstances between the later speech and the current one. Perseverations involve echoing inappropriately something that has gone before.
- *Priming*: these errors are similar to perseverations, although they usually involve the repetition of prior sounds or actions. Many children's games are based on leading people into error through recurrent sound primes. Here is an example: From what tree does an acorn come? *(Oak.)* What noise does a frog make? *(Croak.)* What do you call a funny story? *(Joke.)* What is another name for a cape? *(Cloak.)* What rises up from a bonfire? *(Smoke.)* What do you call the white of an egg? Here, the vast majority of people will respond with *'yolk'* – it is almost irresistible. Three contextual factors combine to make the error highly likely: the prior phonological priming; the fact that a correct answer to the question (the white) was in the question itself – and that is most unusual; lastly, another correct answer (albumin) is a seldom used word in our vocabulary – 'yolk', on the other hand, is very strongly associated with 'egg'. (This example is repeated in Chapter 5, but in another context.)
- *Interruptions and distractions*: these can result in place-losing errors. When we return to the task after an unscheduled interruption, we may believe that we were further along than we actually were, and thus omit a step; or we can judge ourselves to be not as far along as we actually were, and perform an unnecessary repetition. They also cause errors by 'capturing' the limited attentional resource at some critical point in an action sequence. We shall discuss this at a later point.
- *Stress*: local stressors such as fatigue, fear, heat and noise are neither necessary nor sufficient to cause an error, but there is no question that their presence increases the likelihood of going wrong – more about this later.

Outcome Categories

The vast majority of errors are inconsequential; indeed many pass unnoticed both by the perpetrator and by his or her companions. But in dangerous environments, such as those encountered in high-risk sports and hazardous industries, errors can and do have bad outcomes. In these circumstances, it is usually essential to categorise errors according to the severity of their consequences.

It should be stressed, however, that the upshot of a human error is largely determined by the circumstances rather than by the psychological antecedents. Switching on the kettle rather than the toaster causes amusement and mild embarrassment; manipulating a control wrongly in a nuclear power plant, as at Chernobyl in 1986, can be – and was – catastrophic. By itself this action slip by an operator was not sufficient to cause the reactor explosion, but it formed a necessary part of the concatenation of events that led to the disaster.

There is a strong tendency among managers, journalists and lawyers to see a false symmetry between the causes of an error and its effects. An unintended action that results in one or more fatalities is often presumed to be the product of a monumental blunder when, in reality, it was a commonplace absent-minded slip. It is true, of course, that those professionals who work in potentially risky domains have a 'duty of care' towards their fellow workers and clients. This requires them to be aware of the hazards and to be especially vigilant in circumstances known to provoke error. In short, the greater the danger, the greater is the need for 'error wisdom'.

In general, though, it is a bad mistake to treat errors as a moral issue, even though on some occasions they can be egregious and negligent. Error does not necessarily equate to incompetence – though that has been a strongly held view among health-care professionals, for example.

Fallibility is a part of the human condition. Errors cannot be eradicated, but they can be anticipated and managed accordingly. We can't fundamentally change the human condition, but we can change the conditions under which people work in order to make errors less likely and more easily recoverable.

Error outcomes tend to be graded according to their severity, as shown in the list below:

- *Free lessons*: these are inconsequential unsafe acts that could have had a bad outcome in other circumstances. All such near misses provide an opportunity for learning, either at the individual or the organisational levels.
- *Exceedances*: these are not necessarily errors, although they can be. They are situations in which human performance strays toward the edge of safe limits. Such deviations are the stuff of which bad accidents are made. In commercial aviation, for example, flight data recordings are scanned by computers to identify such things as level busts, excessively fast (or slow) approaches, heavy landings and the like. Similarly, railway systems have long collected and analysed information relating to signals passed at danger (SPADs). British studies showed that a very large proportion of the SPADs collected over a given period were associated with a relatively small number of signals. This suggests that the problem arises less of from SPAD-prone drivers as from the poor conspicuity and siting of certain signals.
- *Incidents*: although the term is widely used, there is no close agreement as to what it entails. In general, incidents are 'close call' events of sufficient severity to warrant reporting and/or internal investigation. They may involve temporary damage or relatively minor financial loss. In health care, for example, they can include events in which minor harm is done to a patient, or where serious harm is only avoided providentially. It is often the case that a serious accident is thwarted by the effective operation of some of the barriers and safeguards, even though some defences may have been bypassed or breached. Analysis of several of these events provides important information as to weak defensive elements. These analyses can also give us an idea of where the 'edge' is between relative safety and disaster. It could be said that incidents act towards accident prevention as inoculation works in preventing illness: a little bit of what could do you harm acts to strengthen the systems defences.
- *Accidents*: these are events with significant adverse consequences: injury, loss of assets, environmental damage and/or fatalities.

> They fall into two quite distinct groups: individual and organisational accidents. The former are high-frequency/low-severity events – slips, trips, falls, bangs and knocks requiring a few days absence from work. These lost-time injuries (or, more exactly, their normalised frequency per N workers over a given period) are often used as an index of an organisation's relative safety and as a means of comparison with other organisations or industries. Organisational accidents, on the other hand, are low-frequency/high-severity events involving explosions, crashes, collapses, releases of toxic material and the like.

Table 3.1 compares the properties of three kinds of event with regard to their volume (frequency), costs (in terms of human, asset and environmental losses), and the amount of contextual information that can be used for identifying 'upstream' contributing factors. It is generally the case that the available contextual information is inversely related to the frequency of the adverse event. Organisational accidents are extensively reported and investigated in contrast to exceedances that are plentiful but very low on explanatory information. Collectively, these events can reveal where the recurrent problems are in relation to location, activity, task and the people involved.

Error Myths

The topic of human error is rich in myths, but here we will focus on just three of them: errors are intrinsically bad; bad people make bad errors; and errors are random and highly variable.

Errors are not intrinsically bad. They are essential for coping with trial-and-error learning in novel situations. They are the debit side of a mental balance sheet that stands very much in

Table 3.1 Comparing the properties of exceedances, incidents and accidents

Types	*Volume*	*Costs*	*Contextual data*
Exceedances	Very high	Very low	Low
Incidents	Moderate to high	Low to moderate	Moderate to high
Accidents	Low to very low	Unacceptably high	Often very high

credit, but each 'asset' carries a penalty. Automaticity, necessary for skills and habitual action sequences, make us prone to actions-not-as-planned (slips). Limited attentional resources, necessary for coherent planned action, leave us prey to inattention and information overload. A long-term memory containing 'mini-theories' rather than bald facts, leaves us liable to tunnel vision and confirmation bias. As mentioned earlier, one of the driving priorities of the human mind is to strive for meaning – we need to make sense of the world in order to function adequately. This is deeply rooted in the human psyche.

A belief common to most children and many adults is the 'just world hypothesis'. This presumes symmetry between mental processes and their outcome. Put simply, it is that bad things happen to bad people, and good things happen to the worthy. But this is not the way of the world – chance and other unforeseeable factors can ruin the best-laid plans. Conversely, good luck can turn a pig's ear or a bad plan into a silk purse.

One of the basic rules of error management is that the best people can make the worst errors. There are many reasons for this. The best people tend to push at the limits of existing practice by trying out new techniques. They are often in a supervisory capacity and are multi-tasking, thus easily distracted or preoccupied. In the maintenance world, for example, managers sometimes elect to carry out 'hands on' tasks ill-advisedly in order to avoid being de-skilled.

Another widespread myth is that errors occur 'out of the blue' and are highly variable in their form. Neither is the case. Errors are not random and they take recurrent and predictable forms. Different errors occur in different situations, as indicated below:

- Errors happen when you know what you are doing – that is while carrying out a routine task in familiar circumstances – but the actions don't go as planned. These errors take the form of systematic 'absent-minded' action slips and memory lapses. They can also appear as trips, fumbles and stumbles.
- Errors can also happen when you think you know what you are doing, as in dealing what appears to be a trained-for problem, but misapply a normally good rule; apply a bad rule; or fail to apply a good rule. These are rule-based mistakes and violations.

- And errors are certain to happen when you encounter a novel situation and are not sure what you are doing. These are knowledge-based mistakes and take a wide variety of forms. These error types will be discussed in more detail later.

Instances of these error types as they might appear, for example, in a medical context are listed below:

- A physician writes a prescription for 5 milligrams instead of 0.5 milligrams (a slip).
- A nurse delivers a dose of medication late (a lapse).
- A physician applies the wrong formula to adjust the dosage of amino-glucoside, an antibiotic drug, to be administered to a patient with renal problems (rule-based mistake).
- A junior doctor fails to make the above adjustment because he/she does not appreciate the requirement for moderating the dose for patients with kidney disease (knowledge-based mistake).

Another indication that errors are not random events is shown by the existence of recurrent error traps, where the same situations keep creating the same kinds of error in different people. I mentioned these earlier and they will be discussed further at various points throughout the remainder of this book.

In the rest of this chapter, I will unpack these three major categories of error and look at the various sub-categories within each one. We begin with 'absent-minded' action slips and memory lapses.

Slips and Lapses

I find it convenient to sub-divide these execution problems into three main types: recognition failures, memory failures and attention failures.

Recognition Failures

These fall into three main categories:

1. *The misidentification of objects, message, signals, and the like:* expectation plays a strong part in these errors. Train drivers, for example, occasionally perceive a red signal aspect as green,

because they have been accustomed to meeting a green aspect at that point. Such errors have had catastrophic consequences (the Harrow train disaster in 1952). Other contributing factors are similarity – in appearance, location, function and the like – between the right and the wrong objects or signals; poor signal-to-noise ratios – indistinctness, poor illumination, ambiguous sensory data – and strong habit – in well-practised and familiar tasks, perceptions become less precise: we sometimes accept a crude match to what is expected, even when it is wrong.

2. *Non-detections*: the failure to detect a signal or problem (a false-negative). Aside from lack of training and inexperience, these errors are more likely under the following conditions: the inspection was interrupted before reaching the defect; the inspection was completed but the individual was preoccupied, tired or in a hurry; the person did not expect to find a problem in that location; one defect is spotted but another, close to it, is missed; access to the task was unsatisfactory.
3. *Wrong detections (false positives)*: this involves wrongly detecting problems or defects that were not actually present. Many systems, however, are designed to be fairly tolerant of false-positives – better to be safe than sorry. However, when this principle is applied in a military defence system, the results can be catastrophic. False alarms play a large and dangerous part in eroding the trust operators have in their warning and alarm systems.

Memory Failures

Slips and lapses can arise at one or more of the following information-processing stages: input in which insufficient attention is given to the to-be-remembered material and it is lost from short-term memory; storage in which the to-be-remembered material decays or suffers interference in long-term memory; and retrieval when known material is not recalled at the required time (e.g., tip-of-the-tongue states).

Input Failures

What are we most likely to forget on being introduced to someone? It is his or her name. Why? Because the name is part of a torrent of new information about this person and often fails to get taken in unless we make a special effort to focus on the name – and then we often forget aspects of their appearance and what they did for a living. This shows, once again, that the right amount of attention is an important precondition to being able to remember it later.

A second kind of input failure is the forgetting of previous actions. Again, this is due to a failure of attention: the information simply wasn't encoded. When we are doing very familiar and routine tasks, our conscious minds are almost always on something other than the job in hand. This relative inattention is a necessary feature for the task to be done smoothly; conscious 'interrogations' of habitual actions disrupt. For example, it would be unwise to concentrate on what your feet were doing when running down stairs two at a time. Because our minds are on other things we 'forget' where we put things down, or find ourselves walking around looking for something that we are still carrying.

Another consequence of this kind of forgetting is losing our place in a familiar series of actions – we 'wake up' and don't know immediately where we are in the sequence (see earlier). And there is also the 'time-gap' experience where we can't remember where we have been driving or walking in the past few minutes, or what exactly we have been doing. For example, we can be showering and not remember whether or not we have put shampoo onto our hair. The evidence (if there was any) has been washed away while our mind was 'absent' from the details of the task.

Storage Failures

Perhaps the commonest of these is forgetting intentions. An intention to do something is rarely put into action immediately. Usually it has to be held in memory until the right time and place for its execution. Memory for intentions is called prospective memory, and it is particularly prone to forgetting or sidetracking, so that the action is not carried out as intended.

It is, of course, possible to forget an intention so that no trace of it remains. More usually, however the forgetting occurs in degrees. These different levels of forgetting are listed below:

- *Forgetting the plan*: almost complete forgetting leads to the vague 'I-should-be-doing-something' feeling. Here you have an uneasy sense that you should be carrying some action, but can't remember what or where or when it should be done.

- *The 'what-am-I-doing-here?' feeling*: This is a fairly common experience when you initially remember the intention and start to carry it out, but somewhere along the line (usually because you are preoccupied with something else or are distracted) you forget what it is you came to some place to do. You can find yourself looking into an open drawer or refrigerator, or standing at a shop counter, and your mind is a blank. The intention has been lost – although you can be reminded of it shortly afterwards.
- *Forgetting items in a plan*: here you set out to perform a plan of action, think that you've completed it, but later discover you've left something out. A common experience is to return home to find a letter you had meant to post still on the hall table. It is also my experience that it is quite easy to forget things that other people have asked you to do.

Retrieval Failures

This can take very embarrassing forms as in trying to introduce a person whose name you know well, but at the moment of saying it your mind is a blank. It's the 'this-is – er' experience. As a lecturer, I have frequently said something like 'I have three points to make', and then find that I can't recall the third (or even the second) point. We have already discussed the more private tip-of-the-tongue (TOT) experience. Our studies showed that these TOT states can last quite a long time, but in the end they get resolved in one of three ways: the lost word or name appears as the result of a deliberate search, usually the last of many attempts; the searched for item can pop into your mind out of the blue, usually when you are doing some routine job like washing up or vacuuming; and it could be that a TV programme or newspaper or some other external source mentions the word or name and you recognise it as the item you have been searching. Each of these three ways of concluding a TOT state is equally likely.

There can be little doubt that retrieval failures occur more commonly as you grow older – unlike other slips and lapses which surprisingly occur more frequently in the young, perhaps because we rely more and more on memory aids (lists, calendars, post-it notes, knots in handkerchiefs, and the like) as we age.

Nominal aphasia – not being able to say a name that you know you know – is hardly surprising. Names no longer have much in the way of semantic content. Once upon a time Mr Baker was a baker; but it is not usual nowadays. Having a name like 'Reason' is a mixed blessing. One is subjected to all kinds of oft-heard puns, particularly if you happen to be a professor of psychology; the upside, however, is that people tend to remember the name. On balance, the plus side wins out.

Attention Failures

As noted earlier, attention is a limited resource. Direct it at one thing and it is withdrawn from another. When attention is 'captured' by something unrelated to the task in hand, actions often proceed unintentionally along some well-trodden pathway: strong habit intrusions.

Strong habit intrusions Approximately 40 per cent of all absent-minded slips are of this kind. They take the form of intact, well-organized sequences that recognisably belong to some activity other than the one that is currently intended. This other activity is judged as being recently and frequently engaged in, and as sharing similar, locations, movements and objects with the intended actions.

Absent-minded slips are most likely to occur in highly familiar and relatively unchanging surroundings – kitchens, bathrooms, bedrooms, offices and the like – and during the performance of well-practised tasks that were rated as being recently and frequently performed, and largely automatic in their execution.

Another factor is the inappropriate deployment of the limited attentional resources at some critical choice point in the action sequence. For the most part this involves attentional capture by external distraction or internal preoccupation. But there are occasions when too much attention is directed at some largely automatic action sequence. This usually involves a 'where am I?' query following an interruption. Two wrong answers can ensue: either that I was not as far along as I actually was – resulting in a repetition – or that I was further along – resulting in an omission.

In addition to the general disposing conditions mentioned above, there are at least four more situations in which strong habit intrusions are likely to occur:

1. When a change of goal demands a departure from some well-established routine.
2. When changed local conditions require a modification of some familiar and oft-performed action sequence.
3. When a familiar environment associated with a particular set of behavioural routines is entered in a reduced state of intentionality. For example, we could stray into the bathroom and clean our teeth, even though this was not intended.
4. When features of the present environment contain elements similar or identical to those in highly familiar circumstances. (For example: 'As I approached the turnstile on my way out of the library, I pulled out my wallet as if to pay – although I knew no money was required.')

Strong habit intrusions are extremely widespread, and sometimes disastrous. In the summer of 1982, a double-decker bus on a country route in Wales sheared off its top deck when attempting to pass under a low railway bridge, killing six people. At the coroner's inquest, the driver said: 'It was not in my mind that I was driving a double-decker bus.' He had taken a route he normally drove with a single-decker bus. Other strong habit intrusions have been implicated in the Ludlow (1956) and Lewisham (1957) train crashes, and the calamitous runway collision at Tenerife in 1977.

Interference errors Interference errors result from 'crosstalk' between two currently active tasks (blends and spoonerisms), or between elements of the same task (reversals or spoonerisms). A typical blend is when elements from the previous task carry over into the next. (For example: 'I had just finished talking on the phone when my secretary ushered in some visitors. I got up from behind the desk and walked to greet them with my hand outstretched saying "Smith speaking".') A reversal is when the actions are correct, but the objects for which they were intended get transposed. Consider the following account published in the *Spectator* of 1711:[1]

> My friend Will Honeycombe is one of the Sort of men who are very often absent in conversation . . . A little before our Club-time last night we were walking together in Somerset Garden, where Will picked up a Pebble of so odd a make, that he said he would present it to a Friend of his. After we had walked some time, I made a full stop with my Face towards the West, which Will knowing this to be my usual method of asking what's o'Clock in an Afternoon, immediately pulled out his Watch and told me we had seven Minutes good. We took a turn or two more, when, to my great Surprise, I saw him fling away his Watch a considerable way into the Thames and with great Sedateness in his Looks put the Pebble, he had before found, in his Fob. As I have naturally an aversion to much Speaking, and do not love to be the Messenger of ill News, especially when it comes too late to be useful, I left him to be convinced of his mistake in due time and continued my Walk. . . .

There is a close resemblance between action slips and the errors we find in other domains of mental function, particularly speech. The similarity between the behavioural spoonerisms and slips of the tongue is obvious. Will Honeycombe's error was clearly of the same kind as those attributed to the Reverend W.A. Spooner who is purported to have said such things as 'queer old Dean' when he meant to say 'dear old Queen'.

Other very similar errors also occur in both speech and action; for example, the 'premature exits' from action sequences are closely comparable to the familiar actor's error of being triggered unwittingly into speaking lines from the second act by a similar combination of words in a first act speech. These anticipatory errors also happen quite frequently in everyday speaking and writing.

Let me conclude this section on slips and lapses by summarising three general factors that are involved in promoting these absent-minded errors:

- The *performance of a routine habitual task in familiar surroundings.* Paradoxically, absent-mindedness is the penalty we pay for being skilled; that is, for being able run off our routine actions in a largely automatic fashion.

1 Bond, D.F. (1965) *The Spectator Vol. I.* Oxford: Clarendon Press (pp. 329–330).

- *Attentional capture by preoccupation or distraction.* This capture happens when almost all of the limited attentional resource is devoted to one thing. If it is an internal worry, we call it preoccupation; if it is something happening in our immediate vicinity, we call it distraction.
- *Change, either in the plan of action or in the surroundings.* If no change had occurred, then the actions would have run along their accustomed tracks as intended. Change, of any kind, is a powerful error producer.

Rule-based Mistakes

As stated earlier, human beings are furious pattern matchers. When confronted with an unplanned-for situation we are strongly disposed to identify a familiar pattern and, where necessary, apply a problem-solving rule that is part of our stock of expertise. But these pattern-matching and rule-applying processes can be in error. Rule-based mistakes take three basic forms:

1. We can misapply a normally good rule because we fail to spot the contra-indications. (For example: a GP fails to identify that a child with a fever in a flu epidemic has meningitis.)
2. We can apply a bad rule. (For example: The technician involved in rewiring a signal box just prior to the Clapham rail disaster had acquired the habit of bending back the old wires rather than removing them.)
3. We can fail to apply a good rule. Standard operating procedures (SOPs) usually embody good rules. Failing to comply with SOPs can be both an error and a violation. We will discuss violations in the next chapter.

Knowledge-based Mistakes

Knowledge-based mistakes occur in entirely novel situations when we have run out of pre-packaged problem-solving rules and have to find a solution 'on the hoof'. These are highly error-provoking conditions; indeed, it is usually only trial-and-error learning that leads us eventually to an answer. The errors act like runway markers to mark out the scope of allowable forward progress.

Mistakes at both the rule-based and the knowledge-based levels are shaped by a variety of biases. Here are some of them:

- *Similarity bias*: far from being random, errors tend to take forms that correspond to salient aspects of the problem configuration. Confirmation bias is the product of both similarity bias and bounded rationality (see below) during problem solving.
- *Frequency bias*: when cognitive operations are under-specified (see below), they tend to take contextually appropriate, high frequency forms.
- *Bounded rationality*: the conscious workspace is extremely limited in its capacity. This makes it liable to 'spillage' and overload.
- *Reluctant rationality*: The principle of 'least effort' acts to minimise cognitive strain. This means that we have a strong preference for automatic, parallel processing, even when the conditions demand computationally powerful but effortful serial processing. We are not always aware of the extent to which we employ these unconscious processes in lieu of conscious thinking.
- *Irrationality*: this is an over-used explanation of mistakes, but there can be little doubt that group dynamics can introduce genuine irrationality into the planning process. What could be more irrational than the wilful suppression of knowledge indicating that a certain course of action will lead to disaster?

Conclusion: A General Rule

If there is one principle that governs the shape of nearly all types of human error, slips, lapses and mistakes, it is *under-specification*. Errors arise when the mental processes necessary for correct performance are incompletely specified. These under-specifications take many forms: inattention, forgetting, incomplete knowledge, ambiguous sensory data, and the like. Fortunately, although under-specification can take many forms, the mind's response is very predictable. It 'defaults' to a response that is frequent, familiar and appropriate for the context. This is very adaptive. When in doubt, our mental processes resort to a response that has proved itself to be useful under these particular circumstances – and that means that it is something that is frequently (and often recently) employed in this context. This is psycho-logic: it may not be correct, but it is very sensible when one is forced to guess.

Chapter 4

Violations and the Varieties of Rule-related Behaviour

Chernobyl and Zeebrugge

It was the Chernobyl disaster in April 1986 that first aroused my interest in violations.[1] This was largely due to human actions: mistakes on the part of the experimental planners, one serious operator slip (undershooting the required power level), and a series of ill-judged but deliberate deviations from safe operating procedures just prior to the explosions. It was these last unsafe acts that appeared to require a distinction between errors and violations.

The operators' actions in the last half hour, though involving major transgressions of plant-operating procedure, were all consistent with their goal of achieving the conditions required for the repeated testing of an electrical device. Ironically, this voltage generator was designed as a safety measure. In the event of an off-site power failure, it was intended to bridge the two to three minute gap before the standby diesel generators could produce sufficient power to drive the pumps of the emergency core cooling system.

Procedural violations may be committed for many reasons. Usually, they are deliberate but non-malevolent deviations from safety procedures, rules and regulations. While the non-compliance is intended, the bad outcomes that occasionally ensue are not – unless, that is, the violations are committed by terrorists or saboteurs. We will consider the many different kinds of violation later, but for now let us return to the Chernobyl tragedy.

---

1 Reason, J. (1967) 'The Chernobyl errors'. Bulletin of the *British Psychological Society*, 40: 201–206.

The power plant operators were caught in a system double bind. They were given a task that was not only beyond their experience and competence, but which made violations inevitable. Some of these were written into the plan – disconnecting the emergency core cooling system, for example – others were necessary to allow the electrical engineers from Moscow the opportunity of repeated testing: uncoupling the steam drum and the turbine automatic safety systems. As in many other disasters, the unfortunate operators were the inheritors of a complex series of failures in the system at large. If we are to understand the nature of violations, we have to look beyond the actions of the people on the spot and examine the weaknesses of the total system.

Evidence heard at the Zeebrugge disaster (occurring in March 1987) suggests that a comparable system double-bind existed aboard the roll-on-roll-off ferry, the *Herald of Free Enterprise*[2]. The ferry sailed from Zeebrugge with the bow doors open in contravention of shipping regulations. The ship capsized soon after leaving the harbour when water entered the open mouth of the car deck, toppling its intrinsically unstable (top-heavy) design. Why were the doors not shut? Because there were only two officers available to supervise three widely spaced operations, despite earlier industrial action to achieve adequate crewing. Why was the Master not aware that the doors were open? Because the ship operated a system of negative reporting: in the absence of any message to the contrary, he assumed the doors had been shut. Furthermore he had no direct way of knowing that they were open or shut. The management had earlier refused a request to install warning lights on the bridge on the grounds of expense. These were subsequently fitted to other ferries in the fleet for a few hundred pounds apiece.

'Impossible' Accidents

The Chernobyl operators and the crew of the *Herald* would surely not have committed their respective violations had they believed that their actions would lead to their disastrous consequences. So why did they act as if they were invulnerable?

---

2 Sheen, Mr Justice. (1987) *MV Herald of Free Enterprise*. Report of Court No. 8074 Fomal Investigation. London: Department of Transport.

Each group had probably committed violations with impunity in the past. Only when these particular violations combined with a large number of other factors did they lead to catastrophe; and in neither case was any one person in a position to predict the future conjunction of these singly insufficient but necessary causal strands – a situation that Willem Wagenaar[3] has aptly termed 'the impossible accident'. In addition, the possibility of a catastrophic outcome, by virtue of its rarity, would not have weighed heavily when set against the advantages gained by achieving the immediate production goals. It may even be that such consequences were unimaginable to those at the sharp end, and thus discounted altogether.

Violations Considered as Unsafe Acts

As we shall see shortly, violations come in a variety of forms and not all of them, by any means, are unsafe. However, at the outset of our research into violations – that is, studies relating to road traffic accidents and lost time injuries in oil and gas exploration – the emphasis was very much upon their dangers.

We grouped errors and violations under the general heading of unsafe acts, and classified non-compliances (like errors) according to the level of performance at which they occurred: skill-based, rule-based and knowledge-based. In each case the decision not to abide by the rules and procedures was shaped by individual, contextual, social and systemic factors, though the balance of these influences varies from one type of violation to another.

Violations at the Skill-based Level

These violations form part of a person's repertoire of skilled or habitual actions. They often involve corner-cutting (i.e., following the path of least effort between two task-related points). Such *routine violations* are promoted by inelegant procedures and

3 Wagenaar, W.A. (1986) *The Cause of Impossible Accidents*. The Sixth Duijker Lecture, University of Amsterdam. See also Wagenaar, W.A. and Groeneweg, J. (1987) 'Accidents at sea: Multiple causes and impossible consequences'. *International Journal of Man-Machine Studies*, 27: 587–598.

a relatively indifferent environment. That is, one that rarely punishes violations or rewards compliance.

Looking down on a city park, for instance, we can see the walking routes that were intended by the landscape architect. And we can also see the muddy tracks through the grass that the park's users preferred if they intended to take the shortest path between, say, a bus stop and an underground station.

In general, there is little need to proceduralise activities at the skill-based level. For the most part, actions are governed by stored habits of action whose details are, in any case, beyond verbal control, or even recall. There is no point, for example, in writing procedures to tell a skilled tradesperson how to use a screwdriver.

Where procedures do cover activities at the skill-based level, they tend to take the form of general exhortations (e.g., Proceed with due caution ... Care should taken when ... etc.).

Optimising violations – thrill-seeking violations – also feature large at the skill-based level. This category is not so much a separate type of violation as an acknowledgement that human action serves a variety of goals, and that some of these are unrelated to the purely functional aspects of the task. Thus a driver's functional goal is to get from A to B, but in the process he or she (usually he) can seek to optimise the joy of speed or indulge aggressive instincts. Similarly, mariners may deviate from safe operating procedures in order to alleviate the tedium of an otherwise uneventful voyage – for example, they can sail closer to an approaching vessel in order to demonstrate their ship-handing skills (a contributing factor in several collisions).

These tendencies to optimise personal rather than strictly functional goals can become an embedded part of an individual's performance style. We see this very clearly in car driving. They are also characteristic of particular demographic groups, most particularly young males.

Violations at the Rule-based Level

Safety procedures, rules and regulations are written primarily to control behaviour in problematic or risky situations, and are most abundant at the rule-based level of human performance.

In the initial stages of a particular system or technology, the procedures may simply provide instructions on how to do the job and how to deal with foreseeable hazards. But procedures are continuously being amended to incorporate the lessons learned from past incidents and accidents. Such modifications usually prohibit specific actions that have been implicated in some previous adverse event. The upshot is that the range of allowable actions gradually diminishes as the technology matures. However, the range of actions necessary to complete the task within operational or commercial constraints may not diminish. In short, the scope of allowable action becomes less than the scope of necessary action. Whereas errors arise from the under-specification of mental processes (see Chapter 3), violations can be provoked by regulatory and systemic over-specification of permitted actions. This creates the conditions for *necessary or situational violations.* These are situations for which violations offer possible or, in some cases (e.g., Chernobyl), the only solutions.

The character of situational violations can be illustrated with an example drawn from railway shunting. The British Rail 'Rule Book' (amended every six months) prohibited shunters from remaining between wagons during easing up; that is, when a set of wagons is propelled by a pilot engine towards some stationary wagons to which they will be attached. Only when the wagons are stopped can the shunter compliantly get down between them to make the necessary coupling. On some occasions, however, the shackle for connecting the wagons is too short to be coupled when the buffers are at their full extension. The job can only be done when they are momentarily compressed as the wagons first come into contact. Thus, the only immediate way to join these particular wagons is by remaining between them during the easing-up process.

In the last days of British Rail (prior to 1994), an unacceptable number of shunters died each year as the result of being trapped between the buffers. This illustrates a sad point: violations per se need not harm you; it is the errors that are made when violating that can prove fatal. We will return to this point later.

The shunting example illustrates an important point about situational violations: whereas routine and optimising violations are clearly linked to the attainment of personal goals – that is, least effort and thrills – necessary violations have their origins in the deficiencies of the workplace and system. Initially, non-compliance is seen as essential in order to get the job done. But, once done, it is often seen that they are an easier way of working and become part of the person's habitual skill-based performance.

Rule-based violations are likely to be more deliberate than skill-based violations. However, just as mistakes are intentional actions carried out in the belief that they will achieve their desired ends, so situational violations are deliberate acts carried out in the belief that they will not result in bad consequences. These violations are shaped by cost-benefit trade-offs (of which more later) where the benefits are seen as outweighing the possible costs.

Such assessments can be mistaken. Thus, situational violations can involve both mistakes and procedural non-compliances. We will discuss 'misventions' (a blend of mistaken circumventions) further when we come on to consider the varieties of rule-related behaviour.

Violations at the Knowledge-based Level

Activities at the knowledge-based level take place in atypical or novel circumstances for which there is unlikely to be any specific training or procedural guidance. Trainers and procedure writers can only address known or foreseeable situations.

The Chernobyl disaster provides perhaps the best-documented account of *exceptional violations*. When the power level fell below 25 per cent, the plant was in a dangerous condition, being liable to positive void coefficient (reactivity that could and did spiral out of control). After that, almost all the activities were exceptional violations – or, more accurately, mistaken circumventions – that made the explosions inevitable. The operators persisted in successively shutting down safety systems in apparent ignorance of the basic physics of the reactor and in the hope of completing the tests in a diminishing window of opportunity.

Problems encountered at the knowledge-based level do not have to be novel in the sense that the surface of Mars would be to some future astronaut. Quite often they involve the unexpected occurrence of a rare but trained-for situation, or an unlikely combination of individually familiar circumstances.

Consider the following situation (one that has occurred on a number of occasions with fatal consequences): two people are inspecting an oil pipeline. One of them jumps into the inspection pit and is overcome by deadly hydrogen sulphide fumes. His colleague, although trained to cope with such a situation (i.e., to radio for help and stay out of the pit), obeys a primitive impulse and jumps down to help his partner, whereupon he too is overcome. The problem, although covered by training, had never been met before by this person in its harsh reality.

This is an area in which violations can shade into heroic recoveries. We will discuss these in a later chapter.

Who is Most Likely to Violate?

Whereas error rates do not change markedly with gender or age (at least within the normal span of working life), those people most likely to bend the rules form a comparatively easily identified group. Their principal demographic and psychological characteristics are listed below:

- Young men;
- Having a high opinion of their work skills relative to others;
- Who may be relatively experienced and not especially error prone;
- Who are more likely to have a history of incidents and accidents;
- And who are significantly less constrained by what other people think and by negative beliefs about outcomes.

Why Do People Violate Safety Rules?

Given the close association between violating and young males, it is tempting to put it all down to an excess of testosterone. Young males generally have rude health, quick reflexes and are at the peak of their physical abilities, all of which cry out to be tested to the limit.

Fortunately, the psychological and physical pressures to violate diminish fairly rapidly with advancing years. In part, this increased compliance is associated with a growing awareness of one's own mortality, morbidity, vulnerability and general frailty in the face of dangerous hazards, not to mention increased responsibilities and family ties.

Equally, or perhaps even more importantly, age-related compliance is also due to the middle-aged and the elderly having different reference groups (people whose opinions they value) to the young, and these more mature 'significant others' don't, in general, condone violations. The same factor probably also plays a large part in gender differences: violating is not something that other women are especially likely to value or admire.

In any case, putting it all down to testosterone doesn't get us very far, since we have no socially acceptable way of doing anything about it. (Indeed, most societies exploit these young male attributes by recruiting them to fight their wars and police their streets.) So we have to ask more manageable questions. In particular, what are the attitudes, beliefs, group norms and situational factors that promote potentially unsafe violations? Some of these can be changed, albeit with some difficulty.

Our research on driving violations suggests that non-compliance is directly related to a number of potentially dangerous beliefs. Some of the more important of these 'illusions' are listed below:

- *Illusion of control:* habitual violators feel powerful and overestimate the extent to which they can govern the outcome of risky situation. Paradoxically, they can also have the opposite feelings (powerlessness) in certain circumstances – such as speeding – where they feel that their own behaviour is merely conforming to the local traffic norms.
- *Illusion of invulnerability:* violators underestimate the chances that their rule breaking will lead to adverse consequences. Skill, they believe, will always overcome hazards. By the same token, young men do not see themselves as the likely victims of other people's bad behaviour. In a recent study, young males were asked to judge the likelihood that they would be victims of street crimes relative to other demographic groups. They made a sevenfold underestimate

of their actual chances of being mugged or assaulted on the streets. They are, in fact, the demographic group at greatest risk. Similar tendencies are likely to operate when driving or working in hazardous conditions. They feel 'fireproof'.
- *Illusion of superiority:* this has two aspects. First, people who score highly on self-reported violation questionnaires rate themselves as being more skilled (particularly in driving) than others. Second, they do not view their own tendencies to violate as being worse than those of other people.

We can also express these tendencies to violate as a series of statements:

- 'I can handle it.'
- 'I can get away with it.'
- 'I can't help it.'
- 'Everyone does it.'
- 'It's what they [the company] really want.'
- 'They'll turn a blind eye.'

The Mental 'Economics' of Violating

Violations are deliberate acts. People can weigh up the perceived costs and benefits of an act of non-compliance, and when the benefits exceed the possible costs they are likely to violate[4]. A table outlining the violation 'balance sheet' is shown in Table 4.1.

For many acts of non-compliance, experience shows that violating is often an easier way of working and brings no obvious bad effects. The benefits are immediate and the costs are seemingly remote and, in the case of accidents, unlikely.

The challenge here is not so much to increase the costs of violating (by stiffer penalties and the like) but to increase the perceived benefits of compliance. That means having procedures that are workable and describe the quickest and most efficient ways of performing the task. Any lack of trust caused by inappropriate or clumsy procedures will increase the perceived benefits of violating.

4 Battmann, W. and Klumb, P. (1993) 'Behavioural economics and compliance with safety regulations.' *Safety Science*, 16: 35–46.

Table 4.1 Summarising the violation 'balance sheet'

Perceived Benefits	Perceived Costs
Easier way of working	Causes accident
Saves time	Injury to self or others
More exciting	Damage to assets
Gets the job done	Costly to repair
Shows skill	Sanctions/punishment
Meets a deadline	Loss of job/promotion
Looks macho	Disapproval of friends

Bad Procedures

It would be a mistake to think that most violations are due to bloody-mindedness on the part of the workforce. Attitudes and beliefs leading to non-compliance are only half the problem. The other half, or more, arises from bad procedures.

In the nuclear industry, for example, nearly 70 per cent of all human performance problems could be traced to bad procedures. That is, procedures that gave the wrong information, or were inappropriate and unworkable in the current situation, or were not known about, or were out of date, or that could not be found, or could not be understood, or that simply had not been written to cover this particular task. Bad, absent or unworkable documentation is not a monopoly of the nuclear power industry.

Procedure-usage

In a survey of procedure usage within a large petrochemical plant in the north-west of England,[5] it was found that whereas safety-critical and quality-critical jobs are associated with a high proportion of procedure usage (80 per cent), less than half of the respondents stated that they used procedures while solving problems (30 per cent) – even safety-critical ones – or while carrying out maintenance work (10 per cent). However, only 58 per cent of the 4000 people surveyed reported that they had the procedures open and front of them while they are actually

5 Embrey, D.E. (1999) Personal communication.

carrying out jobs. People do not usually do and read at the same time.

In many highly proceduralised industries, it is common for the workforce to write their own accounts of how jobs should be done. These are jealously guarded and passed on to new members of the workgroup. They are often known as 'black books'. The procedure-usage survey, discussed above, found that 56 per cent of operators and 51 per cent of managers used these informal procedures.

Below are listed some of the reasons given by petrochemical workers for not following the formal procedures:

- If followed to the letter, the job wouldn't get done.
- People are not aware that the procedure exists.
- People prefer to rely on their own skills and experience.
- People assume that they know what is in the procedure.

Testing Two Models of Violating Behaviour

Behavioural Cause Model

This model was tested on 182 operators working on an offshore platform in the North Se.[6] It was found that the model allowed successful prediction of 64 per cent of the variance in violating behaviour with just four major factors providing the dominant drive to violate. These are listed below:

1. *Expectation*: the person's estimation of the likelihood that they will perform some specified behaviour (such as violating).
2. *Opportunity*: the possibilities an individual has to work in another (better more efficient) way and their judgement of the consequences of working in that way.
3. *Powerfulness*: the person's feeling of superiority, competence and skill based on their experience.
4. *Planning*: the quality and efficiency of the planning process that precedes the work.

6 Verschuur, W., Hudson, P., and Parker, D. (1996) *Violations of Rules and Procedures: Results of Item Analysis and Tests of the Behavioural Cause Model*. Field Study NAM and Shell Expro Aberdeen. Report Leiden University of SIEP.

Knowing an individual's scores on these four factors allows the prediction of whether they are likely to violate or not. These predictions account for two-thirds of the variance in actual violating behaviour. Predicting or explaining the variance is not only a measure of accuracy, it also indicates how much other factors not considered could also be influential. The fact that these four factors explained 64 per cent of the variance means that any other factors would only have a minority effect.

To put these proportions into perspective, most behavioural predictions are about 20–30 per cent accurate and even large opinion polls can only reach such accuracy with samples larger than 1000 and very simple voting behaviour.

Supervision and Punishment Model

An alternative model is to assume that people are bad and lazy. Violating is the norm unless they are forced to comply by (a) detection of violations by supervisors and (b) strong punishment meted out to the violator on detection. When tested, as above, this model was found to account for only 20 per cent of the variance.

When the two models are combined (by adding both sets of factors into the equation), the total variance explained only rose from 64 per cent to 67 per cent, a negligible increase.

The message is clear. Effective management of potentially dangerous violating behaviour depends upon an understanding of the significant driving factors rather than relying upon untested preconceptions. Focusing upon detection, supervision and punishment will only produce marginal improvements; while concentrating upon the four factors of the Behavioural Cause Model could produce major reduction in violating behaviour.

The Varieties of Rule-related Behaviour

So far, we have looked at violations from the point of view of managers of hazardous industries for whom violations are regarded as a major threat to safety. This is indeed true, but there is a wider perspective that starts from the premise that neither compliance nor violating is intrinsically good or bad – it all depends on the local context. To understand the wider varieties of rule-related behaviour, we need to consider a number of systemic and personal factors.

Rule Quality

Since all the ways in which harm can come to people or assets can never be wholly knowable or considered likely, there will always be situations for which no safety procedures are available. And, as we have seen earlier, the procedures can be wrong or inappropriate for the circumstances. Thus, for any one situation, there can be good rules, bad rules or no rules.

Correct and Incorrect Actions

Here, the extent to which an action may be deemed correct or incorrect depends upon the accuracy of the actor's hazard perception. Recognising that a situation is dangerous or that a particular procedure is inappropriate is likely to lead to correct behaviour – that is actions shaped by an appropriate awareness of the need to minimise the risk. Behaviour that disregards the dangers, even though it may satisfy the individual's personal goals, is likely to be incorrect.

Psychologically Rewarding and Unrewarding Actions

Psychologically rewarding actions are those that satisfy the personal goals of the actor. These can be in line with the organisation's objectives or not; they can be compliant or non-compliant, correct or incorrect. For some people, violating serves a personal need for excitement or least effort; for others, it can be a source of guilt and worry even when the rule is inappropriate for the situation. Table 4.2 shows a summary of the 12 varieties of rule-related behaviour and these are discussed in more detail as follows:

1. *Correct and rewarding compliance* In any moderately successful organisation, this is likely to be the largest single category of rule-related behaviour. Procedures are tweaked and adjusted over the years so that they become a more efficient and safer way of working. If this is how they are perceived

Table 4.2 Summarising the 12 varieties of rule-related behaviour

Where the task was covered by an appropriate rule or procedure (good rules)

- Was the procedure followed and was it psychologically rewarding?
    - If YES → Correct and rewarding compliance (1)
    - If NO → Correct but unrewarding compliance (2)
- If the procedure was not followed was it psychologically rewarding?
    - If YES → Incorrect but rewarding violation (3)
    - If NO → Mistaken circumvention (misvention) (4)
- Was the non-compliance motivated by a desire to damage the system?
    - If YES → Malicious circumvention (malvention or sabotage) (5)

Where the task was covered by some inappropriate rule or procedure (bad rules)

- Was the procedure followed and was it psychologically rewarding?
    - If YES → Incorrect but rewarding compliance (6)
    - If NO → Mistaken compliance (mispliance) (7)
- If the procedure was not followed was it psychologically rewarding?
    - If YES → Correct violation (8)
    - If NO → Correct but unrewarding violation (9)
- Was the compliance motivated by a desire to disrupt the system?
    - If YES → Malicious compliance (malpliance or working-to-rule) (10)

Where the task was not covered by a rule or procedure (no rules)

- Did the knowledge-based improvisation yield a good or acceptable outcome?
    - If YES → Correct improvisation (11)
    - If NO → Mistaken improvisation (12)

by the workforce, then compliance will, in general, be more psychologically rewarding than non-compliance.

2. *Correct but unrewarding compliance* Even in the best organisations, however, there will be situations in which the rules are viewed as necessary but nonetheless irksome. Wearing hard hats, high-visibility garments and safety boots on a hot day can be very trying, even though they are seen as necessary for preserving life and limb. Road works that require traffic to alternate in both directions along a single lane signalled by temporary red and green lights can be very frustrating, particularly when we see a clear way through ahead. Usually, though, we curb our impatience and obey the lights because we accept the need for their flow control and welcome an improved road surface.

3. *Incorrect but rewarding violation* These are dangerous because they are habit-forming. Every incorrect but personally rewarding unsafe act increases the likelihood that it will be repeated over and over again, becoming a part of the individual's routine skill-based activities.

   As mentioned earlier, it is not the violation per se that is necessarily dangerous, but the fact that it can increase the probability of a subsequent error in an unforgiving environment. Driving at 100 mph need not in itself be hazardous, rather it is that the driver can become over-confident about judging speed and distance when the costs of a mistake could be fatal.
4. *Mistaken circumvention (misvention)* These are violations that are neither correct nor rewarding and which carry a high penalty. In these instances, the decision to deviate from appropriate safety rules is almost certainly mistaken. The most tragic example of misventions was the behaviour of the Chernobyl operators discussed earlier.
5. *Malicious circumvention (malvention)* Malventions are rule-breaking actions in which the perpetrators intend that their violations should have damaging consequences. They range from vandalism, often committed by boys in their mid-teens, to gross acts of terrorism like those which occurred in New York and Washington on 9 September 2001 and in London on 7 July 2005, as well as in many other cities of the world. In between, there are crimes such as arson, vehicle ramming and many other forms of malicious harm. For the most part, these bad acts lie outside the scope of this book; but their occurrence in the world of hazardous work cannot be altogether discounted. Vandalism on the railways, for example, remains a significant threat in many parts of Britain.
6. *Incorrect but rewarding compliance* Adherence to inappropriate rules, even when they are recognised as such, can be characteristic of people for whom any kind of non-compliance is a source of considerable personal discomfort. It is not in their nature to bend the rules, good or bad; such deviations are 'more than their job's worth'. The judges at Nuremberg in 1946 had much to say on this kind of behaviour.

7. *Mistaken compliance (mispliance)* A particularly tragic instance of mispliance occurred on the oil and gas platform *Piper Alpha,* on 6 July 1988 following an explosion in the gas line. The emergency procedures required that the platform personnel should muster in the galley area of the accommodation towards the top of the platform. Sadly, this location was directly in the line of the fireball that erupted over an hour after the first explosion. Most of those who complied with these procedures died.
8. *Correct violation* Among those who survived the *Piper Alpha* disaster were the divers who deviated from the mustering instructions and descended to the bottom of the platform where they were able to use rope and a ladder to reach a rescue boat.

   Military history is rich in correct violations – though it is usually only the outcome of a battle that determines the correctness or otherwise of the deviations. Nelson won the Battle of Copenhagen because, among other things, he disregarded an order to disengage (by putting his telescope to his blind eye). The Confederate commander, General Lee, violated a basic rule of war at Chancellorsville – don't split your army in the face a superior force – when he sent General Jackson on a 16-mile flanking march that brought his force up to the far end of the Federal line and took them by surprise.

   Such fortunate violations are often taken as the mark of a great commander. For General Lee, however, such deviations could also be seen as a necessity as well as a mark of greatness since the Federal armies that he met were usually larger and always better equipped than his own. But they were, at least in the early years of the war, poorly led and easily thrown by these unconventional manoeuvres. However, even great generals have their bad days, as Lee did at the Battle of Gettysburg later in 1863. By failing to occupy Cemetery Ridge when it was largely empty of Federal troops, he created the necessity for Pickett's disastrous charge and lost the opportunity to win the war. Washington was only a few miles further on.

9. *Correct but unrewarding violation* Here an individual recognises that the local procedures are inappropriate for the task and, unlike the 'Jobsworth' discussed earlier, he or she elects not to comply with them. Although this is the correct course of action, it does not necessarily dispel his or her sense of unease at not obeying the rules. In this sense, therefore, the violations although correct are personally unrewarding.
10. *Malicious compliance (malpliance)* Rigid adherence to rules and procedures – or working to rule – was used quite often in Britain's dispute-ridden railway industry as a weapon in the labour armoury. It was the opposite of illegal, as strikes might have been, but it was nonetheless very effective. Its aim was disruption not damage, which puts malpliance into an altogether different league from malvention. When train crews worked to rule they did not endanger themselves or their passengers. Instead, amongst other things, they insisted on taking all the breaks and rest periods that were due to them, though not regularly claimed. The upshot was that trains were delayed and became scattered all over the country at the end of each day, causing major disturbance to the railway timetable.

    In other industries, working to rule has been used as a tactic of protest, seeking to show management how unworkable, excessively bureaucratic and stifling their rules and regulations were. Some degree of ill-will is present, but it's not of the kind that motivates terrorists, vandals and criminals.
11. Correct improvisation This is knowledge-based processing in the absence of rules or procedures that comes up with a happy outcome. Such improvisations are the stuff that some – but not all – heroic recoveries are made. These are the subject of Chapter 11.
12. Mistaken improvisation Failure to achieve a good outcome in the absence of procedural guidance can be unlucky as well as mistaken. Since knowledge-based problem-solving advances by trial-and-error learning, mistakes are inevitable. The deciding factor is the degree to which the situation is forgiving or unforgiving.

Great Improvisers

What makes a good 'trouble-shooter'? This is a very difficult question that I have wrestled with for many years, and to which a large part of the remainder of this book is devoted. Some people come up trumps on one occasion but not on another. Two teams may be similar in most obvious respects, but one succeeds where the other does not. Even the best people have bad days. It is my impression that the very best trouble-shooters get it right about half the time. The rest of us do much worse.

Although there is no simple answer to the question of what makes an effective improviser, I feel convinced that one of the most important factors is mental preparedness.

Some operators of hazardous technologies have a cast of mind – either as the result of training or arising from an inbuilt tendency to expect the worst, but usually both – that causes them to act out in imagination possible accident scenarios. Some of these come from their knowledge of past events; others involve linking together a combination of possible but as yet unconnected failures. In order to run these scenarios, they stock their minds with the details of past events. They also review incidents and 'free lessons'. Their interest is in the ways these inconsequential close calls could interact to defeat the systems defences. They appreciate that single faults or breakdowns, either human or technical, are generally insufficient to bring down a complex, well-defended system.

Simulators have proved to be invaluable tools in promoting 'requisite imagination'. A number of near-disasters, most notably *Apollo 13*, have been recovered because somebody wondered 'what would happen if' a number of unlikely events combined and then ran these starting conditions on the simulator.

End Piece

The managers of complex and hazardous technologies face a very tough question: how do they control human behaviour so as to minimise the likelihood of unsafe violations without stifling the intelligent wariness necessary to recognise inappropriate procedures and avoid mispliances? The answer must surely lie in how they choose to deploy the variety of systemic controls that are available for shaping the behaviour of its human elements.

These issues have been discussed at length elsewhere.[7] But a brief summary of these controls would be useful here.

Administrative controls have been divided into two groups: process and output controls. But a closer examination shows that they actually locate the ends of a continuum, with one extreme – process control – relying wholly upon direct guidance from centralised management (via rules and procedures), and the other – output control – entailing its relative absence, at least at the level of the frontline operators. Output control, the adjustment of local outputs to match organisational goals, depends primarily on two other modes of systemic control: social or group controls and self or individual controls. It is within these areas that the main remedial possibilities lie.

The immense variety of potentially hazardous situations requires that the governance of safe behaviour is delivered at the level of the individual work group. The key to the success of the German military doctrine of *Auftragssystem* (mission system) lay in the ability of low-level commanders to fulfil organisational goals, with or without specific orders. Translated from military to industrial safety terms, this means selecting and training first-line supervisors to provide on-the-spot action control when safety procedures are either unavailable or inapplicable.

Such a localised system of behavioural guidance makes heavy demands on the personal qualities and skills of the supervisors. An essential qualification for them is a wide 'hands-on' experience the workplace tasks and the conditions under which they are frequently performed. Such supervisors need to be 'sitewise' both to the local productive demands and to the nature of the likely and unlikely hazards. Equally important is a personal authority derived both from the respect of the workforce and the support of management. The latter, in turn, requires that safety ranks high in the list of corporate goals. Top-level commitment to safe working is an essential prerequisite of effective behavioural control.

---

7 Reason, J., Parker, D. and Lawton, B. (1998) 'Organizational controls and safety: The varieties of rule-related behaviour.' *Journal of Occupational and Organizational Psychology*, 71: 289–304.

But not all hazardous activities are carried out in supervised groups. When frontline operators are relatively isolated, the burden of guidance shifts from social to self controls. These demand training in both technical and mental skills. Crucial among the latter are techniques designed to enhance hazard awareness and risk perception. These are measures that promote 'correct' rather than merely successful performance. Whereas an understanding of the limitations of prescriptive process controls is necessary at the organisational level, improved risk appraisal and enhanced 'error wisdom' hold the keys to safer – that is, 'more correct' – performance at the level of the 'sharp end' individual.

Safe and productive work is not necessarily achieved by striving to reduce non-compliant actions willy-nilly. Rather, it comes from developing a portfolio of controls that is best suited to that particular sphere of operations. There is no single across-the-board best package. Controls must be tailored to both the type of activity and the needs of work teams and individuals. It seems likely, however, that those organisations with the widest spread of controls will achieve the best safety results – provided that this variety of measures is informed and supported by an effective culture. Safety culture, the obstacles facing it and the means to socially engineer it, will be the topic of my last chapter in this book.

Chapter 5
Perceptions of Unsafe Acts

In the last two chapters, I lumped together errors and violations under the general heading of 'unsafe acts'. This is not a very good label since it is only the consequences that determine whether an act is unsafe or not. An act need be neither an error nor a violation, yet it can still turn out to be unsafe, and conversely. Accepting this obvious limitation, however, I will stay with the term for the sake of precedent and simplicity. On occasions, unsafe acts are also referred to as active failures in order to distinguish them from latent conditions.

Previously, I focused upon the varieties of error and rule-related behaviour, and upon the psychological, organisational and contextual factors that promote and shape their occurrence. In this chapter, I am less concerned with the acts themselves and those who commit them than with the way they are *perceived* by significant other people. These 'significant others' include the managers of hazardous systems, directors, shareholders, stakeholders, regulators, media commentators, legislative bodies and those whose lives could be or were adversely affected by any bad outcomes (e.g., patients, passengers, customers, consumers, and end-users of all kinds).

A number of different perspectives exist, and not all are mutually exclusive. Each view constitutes a model of why unsafe acts occur and how they impact upon the operations in question. Each model generates its own set of countermeasures and preventative policies. Some of these views are rooted in folk psychology; others have their basis in engineering, epidemiology, and the law and systems theory. Four such perspectives are considered below: the plague model (or defect model), the person model, the legal model and the system model. Discussion of the person model will include an account of the vulnerable system syndrome that it so often engenders; consideration of the system perspective will review a number of system models that have been influential in the safety field.

The chapter concludes by arguing that the extremes of both the person and the system models – the dominant views in safety management – have their limitations. We need to find a balance between the two that continues to promote systemic improvements while, at the same time, giving those who have little chance of changing the system some mental skills (mindfulness) that will help them to avoid error traps and recurrent accident patterns tomorrow rather than at some undetermined time in the future.

The Plague Model

This is less a model than a gut reaction to an epidemiological study[1] that found, *inter alia,* close to 100,000 people dying each year in the United States as the result of 'preventable' medical errors. In a report to the president,[2] the authors began its first section with the heading 'A National Problem of Epidemic Proportion'.

Calling error an 'epidemic' puts it in the same category as AIDS, SARS or the Black Death (*Pasteurella pestis*). A natural step from this perspective is to strive for the removal of error from health care. But, unlike some epidemics, there is no specific remedy for human fallibility.

An allied view is that errors are the product of built-in defects in human cognition. This is a perspective that leads system designers to strive for ever more advanced levels of automation and computerisation in order to keep fallible humans out of the control loop as far as possible.

These are misleading perspectives. The problem arises from confusing error with its occasional bad consequences. That errors can and do have adverse effects leads many people to assume

---

1 Brennan, T.A., Leape, L.L., Laird, N.M. (1991) 'Incidence of adverse events and negligence in hospitalises patients. Results of the Harvard Medical Practice Study I.' *New England Journal of Medicine,* 324: 370–6. Leape, L.L., Brennan, T.A., Laird, N.M., et al. (1991) 'The nature of adverse events in hospitalized patients. Results of the Harvard Medical Practice Study II.' *New England Journal of Medicine,* 324: 337–84.

2 QuIC (2000) *Doing What Counts for Patient Safety*. Summary of the Report of the Quality Interagency Coordination Task Force. Washington DC: QuIC.

that human error is a bad thing. But this is not the case. First, most errors are inconsequential. Second, 'successful' error-free performance can also have harmful effects, as the events of 9/11 tragically demonstrated. Third, errors can have highly beneficial outcomes as in trial-and-error learning – essential for coping with novel problems – and serendipitous discovery. Suppose Alexander Fleming had not left his laboratory window ajar when he went off on his summer holidays.

As described in previous chapters, errors arise from highly adaptive mental processes. Each error tendency is part of some essential cognitive activity. The ability to automatise our routine actions leaves us prey to absent-minded slips of action. But continuous 'present-mindedness' would be insupportable; we would waste hours each day deciding how to tie our shoelaces. The resource limitations of the conscious workspace make us liable to data spillage and informational overload; but this selectivity is essential for the focused execution of planned actions. A long-term memory that contains 'mini-theories' (knowledge structures or schemas) makes us susceptible to confirmation bias and tunnel vision; but it allows us to make sense of the world – consciousness abhors a vacuum.

Errors per se are not bad. They are a natural consequence of a highly adaptive cognitive system. Correct performance and error are two sides of the same coin. As Ernst Mach[3] put it: 'Knowledge and error flow from the same sources, only success can tell one from the other.' However actions do not have to be successful to be useful; each failure is an opportunity for learning.

The Person Model

If it has a disciplinary basis, the person model is derived from the occupational health and safety approach to industrial accidents;[4] but it is also deeply rooted in folk psychology. Its primary emphases are upon individual unsafe acts and personal injury

---

3 Mach, E. (1905) *Knowledge and Error.* Dordrecht: Reidel Publishing Company. [English translation, 1976].

4 Lucas, D.A. (1992) 'Understanding the human factor in disasters.' *Interdisciplinary Science Reviews*, 17, 185–190

accidents – though it is frequently and inappropriately applied to organisational accidents.

People are seen as free agents, capable of choosing between safe and unsafe modes of behaviour. Unsafe acts are thought of as arising mainly from wayward mental processes: forgetfulness, inattention, distraction, preoccupation, carelessness, poor motivation, inadequate knowledge, skills and experience and, on occasions, culpable negligence or even recklessness.

When I first started working in hazardous technologies back in the mid-1980s, the principal applications of the person model were in those domains where people work in close proximity to hazards – building sites, oil and gas platforms, mining and the like. It was also the case that individuals were likely to be both the agents and the most probable victims of accidents.

More recently, however, this perception of unsafe acts has received a significant boost from the patient safety movement, which had its wider origins in the late 1990s (triggered by high-level reports like that of QuIC, mentioned earlier, and the highly influential Institute of Medicine report)[5]. Healthcare professionals, particularly doctors, are raised in a culture of trained perfectibility. After a long, arduous and expensive education, they are expected to get it right. Errors equate to incompetence, and fallible doctors are commonly stigmatised and marginalised.

Not surprisingly, therefore, the main countermeasures derived from the person model are aimed at influencing those individual attitudinal and cognitive processes that are seen as the main sources of unsafe acts. They include 'fear appeal' poster campaigns, rewards and punishments (mostly the latter), unsafe act auditing, writing another procedure (to proscribe the specific unsafe acts implicated in the last adverse event), retraining, naming, blaming and shaming.

On the face of it, the person model has much to commend it. It is intuitively appealing (and still remains the dominant perception of unsafe acts). Seeking as far as possible to uncouple an individual's unsafe acts from any organisational responsibility is clearly in the interests of the system managers and share

5 Kohn, L.T., Corrigan, J.M., and Donaldson, M.S. (2000) *To Err is Human*. Institute of Medicine. Washington DC: National Academy Press.

holders. And blaming individuals is also legally more convenient, at least in Britain – where a number of attempts to prosecute top management have been thrown out by the judges.

Nonetheless, the shortcomings of the person model greatly outweigh its advantages, particularly in the understanding and prevention of organisational accidents. The heart of the problem is that the person model is inextricably linked to a blame culture. The pathologies of such a culture, collectively termed the 'vulnerable system syndrome',[6] are discussed below.

Vulnerable System Syndrome

In all complex, well-defended systems, a bad event requires some assistance from chance in order to create a path of accident opportunity through the various barriers, safeguards and controls. Notwithstanding this chance element, however, the analyses of many disasters in a wide range of systems suggest that there is a recurrent cluster of organisational pathologies that render some systems more vulnerable to adverse events than others. This has been labelled the 'vulnerable system syndrome' (VSS).

Three pathological entities lie at the heart of VSS: blame, denial and the single-minded and blinkered pursuit of the wrong kind of excellence – the latter usually takes the form of seeking to achieve specific performance targets. These targets may be imposed by senior management or by external agencies such as regulators and government departments. Each of these core pathologies interacts with and potentiates the other two so that, collectively, they form a self-sustaining cycle that will continue to impede any safety management programme that does not have as its first step a determined effort to eradicate, or at least moderate, their malignant influence.

Defences-in-depth play a major part in making major disasters highly unlikely in domains such as commercial aviation and nuclear power generation. But all defences carry a cost. Although diverse and redundant defences protect the system

---

6 Reason, J., Carthey, J., and De Leval, M. (2001) 'Diagnosing "vulnerable system syndrome": An essential prerequisite to effective risk management.' *Quality in Health Care,* 10 (Suppl 2): ii21–ii25.

from foreseeable hazards, they also increase its complexity and make it more opaque to the people who work in it. This lack of transparency is further compounded by the influence of VSS. Together, they conspire to ensure that those whose business it is to manage system safety will often have their eyes firmly fixed on the wrong ball.

Blame

Of the three core pathologies, blaming – or an excessive adherence to the person model (in its human as hazard mode) – is the most tenacious and the most harmful. It is driven by a collection of powerful psychological and organisational processes. They are summarised below:

- *Pointing the accusing finger* is something we all like to do. Not only does it satisfy primitive extrapunitive emotions, it also distances us from the guilty party and endorses our own feelings of righteousness. This is even more the case when we can throw in an 'I-told-you-so' comment as well.
- The *fundamental attribution error* is one of the main reasons why people are so ready to accept human error as an explanation rather than something that needs explaining. When we see or hear of someone performing less than adequately, we attribute this to the person's character or ability. We say that he or she was careless, silly, stupid, incompetent, irresponsible, reckless or thoughtless. But if you were to ask the person why he or she did it, they would almost certainly tell you how the circumstances forced them to behave in the way that they did. The truth, of course, lies somewhere in between.
- Why are we so inclined to blame people rather than situations? It has a good deal to do with the *illusion of free will*. It is this that makes the attribution error so fundamental to human nature. People, especially in western cultures, place great value in the belief that they are free agents, the controllers of their own destinies. We can even become mentally ill when deprived of this sense of personal freedom. Feeling ourselves to be capable of choice naturally leads us to impute this autonomy to others. They are also seen as free agents, able to choose

between right and wrong and between correct and erroneous courses of action.

- When people are given accident reports and asked to judge which causal factors were the most avoidable, they almost invariably pick out the human actions. They are seen as far less constrained than situational or operational causes. Of course people can behave carelessly and stupidly. We all do at some time or another. But a stupid or careless act does not necessarily make a stupid or careless person. Everyone is capable of a wide range of actions, sometimes inspired, sometimes silly, but mostly somewhere in between. Human behaviour emerges from an interaction between personal and situational factors. Absolute free will is an illusion because our actions are always shaped and limited by local circumstances. This applies to unsafe acts as to all other actions.
- The *just world hypothesis* is the belief shared by most children and many adults that bad things only happen to bad people, and conversely. Those individuals implicated in an adverse event are often seen as bad by virtue of the unhappy outcome.
- *Hindsight bias* – the 'knew-it-all-along' effect – acts as the clincher.[7] There is a universal tendency to perceive past events as somehow more foreseeable and more avoidable than they actually were. Our knowledge of the outcome unconsciously colours our ideas of how and why it occurred. To the retrospective observer all the lines of causality home in on the bad event; but those on the spot, possessed only of foresight, do not see this convergence. Warnings of an impending tragedy are only truly warnings if those involved knew what kind of bad outcome they would have. But this is rarely the case.
- At the organisational level, there are further processes at work to reinforce these tendencies to see the frontline personnel as both the primary cause of the mishaps and as the main target for the subsequent remedial efforts. The first is *the principle of least effort.* It is usually easy to identify the proximal errors of the people at the sharp end and to regard these as the 'cause' of the event. That being so, there is no need to look any further. The second is the *principle of administrative convenience*. By limiting the search

7 Fischhoff, B. (1975) 'Hindsight does not equal foresight: the effect of outcome knowledge on judgement under uncertainty.' *Journal of Experimental Psychology: Human Performance and Perception*, 1: 289–99.

to the unsafe acts of those directly in contact with the system, it is possible to restrict culpability accordingly and thus minimise any organisational responsibility. This reaction is especially compelling when the actions include procedural violations – a view that equates non-compliance with guilt and ignores the fact that any feedforward control device can be inappropriate in certain situations, as we have seen in Chapters 3 and 4.

Despite its obvious appeal, however, there are many penalties associated with the person model and its corollary, blame culture.

- By isolating the sharp-end individuals from the context in which their unsafe acts occurred, the managers of the system fail to identify the local provocative factors (e.g., inadequate training, poor supervision, inappropriate tools and equipment, undermanning, time pressure and the like). Most importantly, though, uncoupling the actor from the situation makes it hard to identify recurrent error traps and repeated accident patterns (see Chapter 6). One of the most important aspects of safety management is to identify error-prone situations – and this depends critically upon an incident reporting system.
- A reporting culture is an essential prerequisite to managing safety effectively. This requires a climate of trust that encourages frontline operators to tell of their close calls, near misses and inconsequential but potentially dangerous unsafe acts. Closely investigated accidents are relatively infrequent; only through the analysis and dissemination of these 'free lessons' can the organisational managers learn how close their operations came to the 'edge'. But a reporting culture and a blame culture cannot co-exist. People will not report their actions if they feel that they are liable to blame and sanctions. Even one such occurrence can destroy hard-won trust.
- The factors being targeted by the person model – momentary inattention, misjudgement, forgetting, misperceptions and the like – are often the last and the least manageable part of a contributing sequence that stretches back in time and up through the levels of the system. While the unsafe acts themselves are frequently unpredictable, the latent conditions that give rise to them are evident before the event. We cannot change the human condition, but we can change the conditions under which people work in order to make unsafe acts less likely and, when they occur, to make them more easily detected and corrected.

- Moralistic measures such as sanctions and exhortations have a very limited effect, as shown in Chapter 3, and they can often do more harm than good.

Despite being ineffective and counter-productive, these person-oriented measures continue to be applied. Lacking any reliable information about the true nature of the dangers, or the manner of their occurrence, those who manage the system feel safe. There may be the occasional bad apple, but the barrel itself – the organisation and its defences – is in good shape. Having identified and 'dealt with' the 'wrongdoers', it is then but a short step to believing that it could not happen here again. And anyone who says differently is a troublemaker. Thus blaming fosters denial, the next stage in the unholy trinity that constitutes the vulnerable system syndrome.

Denial: 'It Couldn't Happen Here'

No statement from the managers of a hazardous system could chill me more than 'it couldn't happen here' – although the claim that 'we have an excellent safety culture' comes very close. Denials and boasts of this kind are the antithesis of mindfulness, a topic that will be considered in detail later. But what kind of 'it' did these claimants have in mind? The event most commonly eliciting these statements was the Chernobyl disaster. There were many others as well, but the aftermath of Chernobyl will be sufficient for us to home in on how these denials were expressed. They all took the form: 'It couldn't happen here because ...'

- *Western reactors have adequate containments.* It is doubtful whether any western containment could have withstood the force of the Chernobyl explosions.
- *Western nuclear plants have more competent management.* Consider the following remarks made by the US Nuclear Regulatory Commissioner, J.K. Asseltine[8] 'the many operating events in the Tennessee Valley Authority's plants [led in 1985] to the collapse

---

8 Besi, A., Mancini, G., and Poucet, A. (1987) *Preliminary Analysis of the Chernobyl Accident*. Technical Report No. 1.87.03 PER 1249. Ispra Establishment, Italy. Commission of European Communities Joint Research Centre.

of the TVA nuclear management structure and the indefinite shutdown of all five of TVA's operating plants.'

- *We, unlike Soviet workers, are not motivated to achieve production goals regardless of the risks to human life.* The *Challenger* and Zeebrugge disasters suggest otherwise.
- *We make a systematic study of human errors in nuclear power plants.* During the Sizewell B Inquiry, a CEGB[9] expert witness admitted that they neither collected nor collated human reliability data in any systematic fashion.
- *The RBMK emergency protection system is not a separate and independent system, as in western reactors. As such it is especially vulnerable to operator errors.* True, but the record shows that safety devices in western reactors are highly vulnerable to maintenance errors. A survey carried out by the Institute of Power Operations (INPO) in Atlanta (funded by US nuclear power utilities) found that errors by maintenance personnel constituted the largest proportion of human performance problems in US nuclear power plants.[10] No safety system, however sophisticated, is proof against a 'common mode' maintenance error. A survey carried out at the Joint Research Centre at Ispra identified 67 events in which safety or control systems were rendered inoperable by human error.[11]
- *Western operators are less error-prone than their Soviet counterparts.* Of the 182 root causes of US and European nuclear power plant events occurring in 1983, 44 per cent were classified as human error.[12]

Here is a statement from Lord Marshall, Chairman of the Central Electricity Generating Board and a passionate spokesman for nuclear power generation. The quotation came from the foreword he wrote for the CEGB's case for building a new pressurised water reactor at Hinkley Point, Somerset (it did not get built for financial reasons). I was present at the inquiry as an expert witness for the Consortium of Opposing Local Authorities – not, I hasten to add, because I was against nuclear power, but

9 Central Electricity Generating Board.

10 INPO (1985) *A Maintenance Analysis of Safety Significant Events*. Atlanta, GA: Institute of Nuclear Power Operations.

11 Besi, Mancini, and Poucet (1987).

12 INPO (1984) *An Analysis of Root Causes in 1983 Significant Event Reports*. Atlanta, GA: Institute of Nuclear Power Operations.

because I did not believe that the CEGB had fully grasped the risks due to latent conditions and organisational accidents. Here are Lord Marshall's bold assertions in the aftermath of the Chernobyl disaster – they still send shivers up my spine:

> At sometime in the future when nuclear power is an accepted and routine method of electricity generation could our operators become complacent or arrogant and deliberately and systematically flout operating rules? Could our system of safety reviews and independent assessment and inspection become so lax or so routine that it failed to identify and correct such a trend? My own judgement is that the overriding importance of ensuring safety is so deeply ingrained in the culture of the nuclear industry that this will not happen in the UK.[13]

While it is true that the power-generation arm of the UK nuclear industry has, despite the problems of ageing plants and the novel complexities of decommissioning, avoided any major disasters, the same cannot be said for the nuclear reprocessing components of the industry. The Irish Sea, into which nuclear waste products continue to be poured, remains the most irradiated stretch of water in the world. Its contaminating effects extend up to the Arctic Ocean and the Bering Sea.

Let me take a pause here and try to explain myself. Looking back over the last few paragraphs, I believe you would be forgiven in thinking that I have been demonising the nuclear industry, blaming and shaming them in much the same way as managers might vilify fallible people on the front line. Yes, the brash denials that Chernobyl, or something like it, couldn't happen here were chilling at the time. But a lot has happened since. I have worked with many hazardous technologies over the past twenty years, and there is no doubt that the risk managers of nuclear power plants are among the very best in the business, particularly in the United States, Sweden and Finland. Furthermore, I am committed to the view that people the world over want to keep their lights on, their bathwater warm and their refrigerators, freezers and air-conditioning systems working. Energy has a price, some of which is monetary and some a degree of risk. However the energy package of the future might look, nuclear power generation will and should be a part of it. But let me come to my point: it is not the existence of these denials that bothers me, but the manner in

---

13 UK Atomic Energy Authority (1987) *The Chernobyl Accident and its Consequences*. London: HMSO.

which they are expressed – though some things have changed in the ensuing years, but not everything.

Nuclear power, unlike other forms of generating energy (rightly or wrongly), suffers from a serious disadvantage: people are terrified of radio activity, something that reaches back to Hiroshima and Nagasaki – but which kept the peace for nearly 50 years in the Cold War. Canada carries this burden of gut-fear more than most. Not only do 12 of its nuclear power plants lie close to Toronto, the densest area of population, but its reactors – the CANDU's (CANada Deuterium Uranium – though its similarity to 'can do' is not accidental) – share with the Chernobyl RBMK reactor the feature that increased boiling in the core increases the power: the positive void coefficient. This differs from most other western reactors in which the power goes down as the boiling increases.

Two years ago, I was invited to talk to the nuclear regulators, the Atomic Energy Control Board in Ottawa, and to a CANDU Owner's Group meeting in Toronto. The agreed topics included a broad sampling of what has gone before in this book. Before leaving for Canada, I checked the internet and found a piece produced by the Atomic Energy of Canada Limited, the manufacturers of the CANDU reactor. It was written in November 1996 and entitled 'Why a Chernobyl-type accident cannot happen in CANDU reactors'. The argument rested primarily on a comparison of the safety systems currently possessed by the CANDU reactor and those available in the Chernobyl RBMK reactor in 1986. These comparisons are summarised in Table 5.1.

On the face of it, the CANDU reactors win hands down. They are clearly much better protected than the RBMK reactor. Should a replay of the Chernobyl event now happen in Canada, it is likely that a CANDU reactor would achieve a safe state long before there was any danger of catastrophic explosions. So why don't these engineering superiorities give me greater comfort? Why do I think that these mechanistic arguments miss the plot? My concerns are listed below:

- The paper discussed above was issued by AECL's Marketing and Sales Department.[14] It was primarily a piece of propaganda

---

14 Richard Feynman, a Nobel Laureate, served on the Presidential Commission into the *Challenger* disaster. Later he wrote: 'For a successful

Table 5.1 Comparing the CANDU and RBMK safety features

Feature	CANDU	RBMK
Containment	Full containment.	Partial containment.
Shutdown systems	1. Two complete systems: absorber rods and liquid injection. 2. Two seconds to be effective. 3. Effective independent of plant state.	1. One mechanism: absorber rods. 2. Ten seconds to be effective. 3. Effectiveness dependent on plant state.
Moderator	Cool heavy water.	Hot solid graphite.

designed to allay the fears of Canadian consumers and foreign buyers of CANDU reactors. Canadian nuclear engineers and operators are very good and I don't believe for a moment that they see these engineering features as presenting the whole safety story. (Would they have asked me over if they did?) But everyone prefers good news to bad, and this along with some 170 years of disaster-free (though sometimes troubled) nuclear operations can combine to take the edge off the 'feral vigilance' necessary to remain prepared for the many ways in which complex socio-technical systems can fail calamitously. The Chernobyl operators did not sufficiently respect the hazards facing them. It was not that they had forgotten to be afraid, they had not learned to be afraid – thanks to a Soviet culture that did not share event reports between plants for fear that they would leak out and frighten the consumers.

- Years of successful operation are not, of themselves, proof of adequate safety. At the time of Three Mile Island in 1979, the United States had amassed 480 years of relatively safe power generation from large reactors, and by 1986, the year of the Chernobyl disaster, the USSR had 270 years of nuclear operating experience. The likelihood of serious accidents cannot be judged from statistics such as these, nor do they justify any belief CANDU plants are either more or less safe than other reactor types. The Atomic Energy Control Board found a significant backlog of required

technology, reality must take precedence over public relations, for Nature cannot be fooled.' Feynman, R.P. (1988) *What Do You Care What Other People Think?: Further Adventures of a Curious Character*. New York: Norton (p. 237).

maintenance, out-of-date operating procedures, incomplete inspections and design deficiencies in operating plants.[15]

- Each year there are a variety of safety significant events at Canadian nuclear power plants. Analyses of these events have shown that human error plays a part in more than 50 per cent of these occurrences (as it does elsewhere). The AECB concluded that since 'both the nature and the probability of human error are difficult to quantify, the probability of serious accidents which are a combination of system failure and incorrect human response is difficult to predict.'[16] Despite these problems, the nuclear regulator concluded that Canadian nuclear power plants are acceptably safe. However, they added an important rider: 'safe' does not mean 'risk free'. It means that, on balance, the benefits of generating electricity with nuclear power plants outweigh the risks.'[17] These assessments were based upon technical evaluations and inspections that the AECB even then judged to be too narrow in their scope and depth.
- Nuclear power plants are complex, tightly-coupled socio-technical systems[18] with a diversity of human elements: managers, operators, technicians, maintenance personnel and programmers. Yet AECL base their denials of another Chernobyl almost entirely upon technical safety features – the same was true of the AECB's judgements of Canadian nuclear safety, but they have acknowledged the insufficiency of this narrow approach. AECL could argue that their claims are justified because engineered safety features can defend against both technical and human failures – where human, in this case, generally means control-room operators in a highly automated plant working under routine full-power conditions (more of which later). But this overlooks the fact that both maintenance personnel and managers play a significant role in the development of safety-significant events.[19]

---

15 Submission to the Treasury Board of Canada by the Atomic Energy Control Board, Ottawa, 16 October 1989.

16 Ibid.

17 Ibid.

18 Perrow, C. (1984) *Normal Accidents: Living with High-risk Technologies*. New York: Basic Books.

19 Lee, J.D. and Vicente, K.J. (1998) 'Safety concerns at Ontario Hydro. The need for safety management through incident analysis and safety assessment.'

- I have already mentioned the critical role played by maintenance and testing errors in the development of accidents. Such errors initiated the near-disasters of *Apollo 13* (see Chapter 9) and Three Mile Island and many more besides[20].
- In 1997, Ontario Hydro decided to shut down seven of its twenty nuclear plants at an estimated cost of eight billion Canadian dollars due to safety concerns. This decision was not prompted by a catastrophic accident nor was it due to a primitive and deficient technology. John Lee of Battelle Seattle Research Center concluded that 'the root cause of all of the problems was insufficient attention to human and socio-organisational factors. Ontario Hydro management seemed to believe that nuclear safety could be maintained by technology alone'.[21]
- Kim Vicente, a human factors expert, conducted field research at one of Ontario Hydro's plants. He identified a number of management shortcomings relating to out-dated safety instrumentation and inadequate alarm systems. The most serious was the condition of the Emergency Core Injection flow meters. These had not functioned properly since the plant was built. They indicated flow when there was none. These and other instrumentation problems Vicente described as 'accidents waiting to happen'. Vicente observed: 'The unjustified satisfaction of the Ontario Hydro management concerning safety stems from a failure to consider the full range of factors affecting safety. Their implicit model of system safety seems to have focused on the technical/engineering system ... A more in-depth safety analysis shows that overall system safety is highly dependent on the workers and the organisational and management infrastructure. Focusing on one part of a complex system provides a misleading estimate of system safety.'[22]

Two problems underlie the AECL's assertions. The first is that although there may be recurrent accident patterns (see Chapter 6), organisational accidents never repeat themselves exactly: there are too many contributing factors. Each one can be

---

Proceedings of Workshop on Human Error, Safety and System Development, (http://www.dcs.gla.ac.uk/~johnsom/papers/seattle_hessd/), (pp. 17–26).

20 See Reason, J. and Hobbs, A. (2003) *Managing Maintenance Error: A Practical Guide.* Aldershot: Ashgate.

21 Ibid, p. 22.

22 Ibid, p. 23.

relatively harmless, familiar even; it is only in combination with other singly insufficient contributions that the sequence achieves its lethal power. Such interactions are usually unforeseen and may even be unforeseeable by any normal means.

At Chernobyl, a group of well-intentioned, highly motivated, prize-winning[23] operators managed to blow up a nuclear reactor of proven design without any help from technical failures – other than those they induced themselves. They did it by an unlikely combination of errors and mistaken violations. Such a concatenation of unsafe acts could happen anywhere at any time, although not in the precise manner of the Chernobyl accident. Their likelihood of occurrence cannot be estimated by any of the normal metrics of probabilistic risk assessment. Why? Because they involved a set of non-linear conjunctions, between actions and components that are characteristic of accident trajectories in complex socio-technical systems.

The second problem is that the comparisons between the CANDU and the RBMK were based upon mechanistic rather than systems thinking. A nuclear power plant is a combination of two systems: an engineered system that, by itself, is amenable to a reductionist mode of analysis, and a human-activity system that requires quite another kind of thinking. It is the interactions with these non-mechanical elements that make the totality a complex adaptive system.

The essential characteristic of the reductionist approach is to break a problem down into its component parts. This division continues until the elements are simple enough to be analysed and understood. In comparing the safety potential of the CANDU and RBMK reactors, the defensive features were reduced to the component parts and then compared on a one-to-one basis. This is logical, linear and readily understood. But it misses a crucial point. Jake Chapman[24] expressed the problem very elegantly:

> 'What if essential features of the entity are embedded not in the components but in their interconnectedness? What if its complexity arises from the ways in which its components actually relate to and interact with one another? The very act of

---

23 The Chernobyl operators earlier received an award for exceeding production targets.

24 Chapman, J. (2004) *System Failure: Why Governments Must Learn to Think Differently*. Second edition. London: Demos, (p. 35). This short pamphlet has been of tremendous assistance in elucidating the subtleties of system thinking and the law of unintended consequences.

simplifying by sub-division loses the interconnections and therefore cannot tackle this aspect of complexity.'

So how can we explain the facts that the greater part of nuclear power plant operations are routine and trouble-free? Some answer to this can be gained from a human factors inspection programme carried out by the US Nuclear Regulatory Commission's Office of Operational Data in 1994. This programme focused on the circumstances surrounding a number of potentially significant operational events in the US nuclear power plants (pressurised and boiling water reactors rather than CANDUs).[25] I reviewed 21 of these inspection reports: ten of these involved successful recoveries, and the remainder were less successful. All events ended safely, but in 11 cases the operators showed confusion and uncertainty as to the causal conditions and made errors in their attempt to restore the plant to a safe state. In contrast, the actions of the successful crews were exemplary and, on occasions, inspired.

An important discriminator between the successful and less successful recoveries was the plant state when the off-normal conditions arose. In the majority of the successful recoveries, the plants were operating at or near full power. This is the 'typical' condition for a nuclear power plant, and it is for this state that most of the emergency operating procedures are written. Nearly all successful recoveries were facilitated by these procedures and their associated training.

The less successful recoveries, on the other hand, occurred mostly during low-power or shutdown conditions. These are normal plant states – necessary for maintenance, repairs, regulatory inspections and refuelling (though not in the case of the CANDU and RBMK reactors) – but they constitute a relatively small part of the total plant history. Few procedures are available to cover emergencies arising in these atypical states. As a result, the operators had to 'wing it' and, on several occasions, made errors that exacerbated the original emergency and delayed recovery. The crews operated outside the rules because no rules were provided for handling emergencies in these still dangerous

25 NUREG/CR-6093 (1994) *An Analysis of Operational Experience during Low Power and Shutdown and a Plan for Assessing Human Reliability Issues*. Washington DC: US Nuclear Regulatory Commission.

plant conditions. As a result they were forced to improvise in situations they did not fully understand. It is no coincidence that the Chernobyl unsafe acts occurred during a planned shutdown.

It is highly probably, therefore, that unpredictable non-linearities are most likely to occur on those relatively infrequent occasions when the plant is in a non-routine state as during shutdowns, maintenance inspections and low power operations; even though these atypical conditions are a normal and planned-for part of a plant's operating life.

It is time to move on from the reductionist thinking of some nuclear plant managers and consider some cultural aspects of denial. Ron Westrum,[26] an American social scientist, classified safety culture into three kinds: generative, bureaucratic (or calculative) and pathological. A major distinguishing feature is the way in which an organisation deals with safety-related information – or, more specifically, it's about how they treat the bearers of bad news.

- *Generative or high-reliability organisations* encourage the upward flow of safety-related information. They reward the messengers, even when they are reporting their own potentially dangerous errors. They share a collective mindfulness of the hazards, respect expertise and are reluctant to simplify interpretations. They expect bad things to happen and work hard to prepare for the unexpected. We will discuss these characteristics further in Chapter 13.
- *Bureaucratic or calculative organisations* – the large majority – occupy the middle ground. They don't necessarily shoot the messenger, but they don't welcome him or her either. Bad news and novel ideas create problems. They tend to be 'by-the-book' organisations that rely heavily on administrative controls to limit performance variation on the part of the workforce. Safety management measures tend to be isolated rather than generalised. They prefer local engineering fixes rather than widespread systemic reforms.

---

26 Westrum R (1992). 'Cultures with requisite imagination', in J.A. Wise, V.D. Hopkin and P. Stager (eds), *Verification and Validation of Complex Systems: Human Factors Issues*. Berlin: Springer-Verlag, (pp. 401–416).

- *Pathological organisations* are inclined to shoot the messenger. They really don't want to know. Whistle-blowers are muzzled, maligned and marginalised. The organisation shirks its safety responsibilities, doing only the bare minimum necessary to avoid prosecution and keep one step ahead of the regulator. It punishes or covers up failures and discourages new ideas. Production and the bottom line are the main driving forces.

If, for whatever reason, a management believes that they are operating a safe system (or they lack sufficient respect for the hazards), they are free to pursue the efficiency and cost-saving targets that feature so prominently in modern industrial life. Getting the numbers right and meeting production targets are what professional managers have been trained for and, not unreasonably, they feel that their performance will be judged primarily by their success in achieving these goals. And their bonuses also depend on it. This opens the way to the single-minded and often blinkered pursuit of excellence.

The Wrong Kind of Excellence

Perhaps the most bizarre case of pursuing the wrong kind of excellence occurred in the Royal Navy of the mid-nineteenth century. With the transitions from sail to steam and from cannon to gun turrets, naval officers were faced with the problem of what to do with the many sailors who were no longer required to handle the sails and man the guns. Their solution was to create the cult of 'bright work' in which captains and their crews vied with one another to produce the shiniest surfaces and the glossiest paintwork.[27] This was not altogether irrational since, for the last eighty-odd years of the nineteenth century, the Royal Navy faced little in way of serious opposition and was mostly at peace. Their primary task became 'showing the flag', for which purpose gleaming ships rather than effective gunnery were judged more suitable. Massive watertight doors were lifted from their hinges and buffed and rubbed until they shone like mirrors. But in the process, they ceased being watertight – a fact that cost

27 Massie, R.K. (1992) *Dreadnought: Britain, Germany and the Coming of the Great War*. London: Jonathan Cape.

them dearly in the loss of all hands on *HMS Camperdown*. And insufficient gunnery training brought about the rapid sinking of a British flotilla in one of the earliest engagements of the First World War off the coast of Chile.

The twentieth-century history of maritime disasters is also studded with dramatic instances of the single-minded but blinkered pursuit of excellence that ended in catastrophe. The best-known examples are the sinking of the *Titanic* in 1912, after being driven at full speed into forecasted icebergs in an attempt to clock up a transatlantic speed record, and the capsize of the *Herald of Free Enterprise* just outside Zeebrugge in 1986. The shore-based managers had made their shareholders very happy in the preceding months by winning the stiff commercial competition for cross-Channel passengers; but in so doing they had fatally eroded the slim safety margins on these already capsize-prone ro-ro ferries.

Similar themes are echoed in the King's Cross Underground fire. Mrs Thatcher, the then prime-minister, had replaced the London Underground management with a team that strove strenuously for the 'leaning and meaning' (meaning as in nasty 'mean') management style that characterised the age. The maintenance of the escalator that was the origin of the horrific fire blast had been out-sourced (what a term of the 1980s!). And, just like the cigarette end that set the whole thing ablaze, the maintenance had fallen through the gaps in the system. This theme recurs in the collapse of Barings Bank and many other disasters of the time.

Even those bureaucratic (calculative) organisations with their eye firmly on the safety ball can still pursue the wrong kind of excellence. There are many companies engaged in hazardous operations that continue to measure their systemic safety by the lost time frequency rates of its various divisions. Unfortunately, this reflection of personal injury rates provides little or no indication of a system's liability to a major hazard disaster. As Andrew Hopkins pointed out: the road to disaster is paved with falling or very low lost time injury rates.[28]

28 Hopkins, A. (1999) *Managing Major Hazards: The Lessons of the Moura Mine Disaster*. St Leonards: Allen & Unwin.

Dietrich Doerner, a psychologist and the winner of Germany's highest science prize, has spent many years studying the strengths and weaknesses of human cognition when managing richly interconnected dynamic systems.[29] His findings throw considerable light on this blinkered pursuit of excellence. When dealing with complex systems, people think in linear sequences – we've been there before with Ontario Hydro. They reason in causal series rather than in causal networks. They are sensitive to the main effects of their actions upon their progress towards and immediate (often numerical) goal, but frequently remain ignorant of their side effects upon the rest of the system. In an interactive tightly coupled system, the knock-on effects of interventions radiate outwards like ripples in a pool. But people can only 'see' their influences within the narrow sector of their current concern. It is also the case that people are not very good at controlling processes that develop in an exponential or non-linear fashion. They repeatedly underestimate their rate of change and are constantly surprised at the outcomes.

The problem is exacerbated by close attention to critical numerical targets. Government policy-makers and the managers of complex systems live by numbers, but they don't always appreciate their limitations when applied to complex adaptive systems. Some examples of the 'law of unintended consequences' are given below:

- The Norwegian government, seeking to reduce the health risks associated with excessive alcohol consumption, instituted a huge price rise in beer, wine and spirits. The effect of this was to create a substantial increase in the manufacture and consumption of home-made 'hooch' – which is much more likely to be injurious to health.
- In 2001, an award of £21 million was made to ambulance services for improved performance. This ended up by costing the London Ambulance Service £1.5 million. The bulk of the award went to services outside the London area whose performance was most in need of improvement. They used the funds to recruit more staff, mainly from the London Ambulance Service who lost 75 more staff than it would have done through normal

29 Doerner, D. (1996) *The Logic of Failure*. New York: Henry Holt & Company.

attrition. The cost of making good the lost staff was £20,000 per person for recruitment and training.[30]

- A target was established for urgent referrals of women with suspected breast cancer. This required that such cases were seen by a specialist consultant within two weeks. Referral come through two routes: one is the mass screening programmes carried out by GPs; the other is by attendance at a breast clinic run by a specialist consultant. GPs are good at identifying cases with standard symptoms, and for these women, the fourteen-day target has improved treatment rates. But for women with less obvious or more subtle symptoms (that GPs are less able to detect), the only available route is by attendance at the breast clinic. However, the time to see the specialist has lengthened due to the increased referral rate from the mass screening programmes. As a consequence, the women with more hidden symptoms wait longer to see the specialist, even though their need for rapid diagnosis and treatment is greatest.[31]

There is a clear message from these and other examples of the 'law of unintended consequences': the single-minded and blinkered pursuit of specific numerical targets can lead to a decrease in overall system performance. This is well understood in systems engineering but not necessarily by the managers of complex, well-defended technologies.

The Legal Model

This is a corollary of the person model, but with strong moral overtones. At the heart of this view is the belief that responsible and highly trained professionals such as doctors, pilots, ship's captains and the like, should not make errors. They have a duty of care.

Those errors that do occur, however, are thought to be rare but sufficient to cause adverse events. Outcomes determine culpability to a large extent. Errors with bad consequences are judged to be negligent or even reckless and deserve punitive

30 Chapman (2004), p. 29.
31 Ibid, p. 57.

sanctions. Consider, for example, the following quotation from an Institute for Safe Medication newsletter:[32]

> In the past few months, three fatal medication errors have captured the news headlines; according to these news reports, criminal investigations are being considered in each case. If the criminal investigations proceed, felony indictments could be levied against some or all of the practitioners involved in these errors.

Far from being rare occurrences, the research evidence shows that errors made by highly trained professionals are frequent, but most are either inconsequential or detected and corrected. An observational study in 1997 carried out 44 hours of continuous observations of pilot errors in flight. The authors then extrapolated these error frequencies to worldwide commercial aviation and estimated that there were around 100 million errors committed in cockpits each year, but the accident records indicate that there are only between 25–30 hull losses each year. Slightly tongue in cheek, the authors concluded that errors were necessary for safe flight.[33]

Another study[34] involving direct observations of over 165 arterial switch operations – correcting the congenital transposition of the aorta and pulmonary arteries in neonates – in 16 UK centres and 21 surgical teams found that, on average, there were seven events (usually arising from errors) in each procedure, of which one was life-threatening and the others were relatively minor irritants that interrupted the surgical flow. Over half the major events were corrected by the surgical team, leaving the fatality odds for that child unchanged (see also Chapter 9).

The switch operation takes surgical teams to the very edge of their cognitive and psychomotor performance envelopes. Errors are unavoidable. The virtuoso surgeons make errors, but they anticipate them and mentally prepare themselves to detect and recover them. However, this compensatory ability can be eroded by the minor events, the greater their number the poorer the outcome. We will discuss these 'heroic recoveries' in Part IV of this book.

---

32 Newsletter, Institute for Safe Medical Practices, 8 March 2007 (Horsham PA).

33 Amalberti, R., and Wioland, L. (1997) 'Human error in aviation.' In H.M. Soekkha (ed.) *Aviation Safety*. Utrecht: VSP, (pp. 91–100).

34 De Leval, M., Carthey, J., Wright, D., Farewell, V., and Reason, J. (2000) 'Human factors and cardiac surgery: a multicentre study.' *Journal of Thoracic Cardiovascular Surgery* 119: 661–672.

These observations demonstrate that error-making is the norm, even among highly accomplished professionals. Training and experience do not eradicate fallibility (although they moderate it); they merely change the nature of the errors that are made and increase the likelihood of successful compensation. Far from being sufficient to bring about bad outcomes, the record shows that only very occasionally are the unwitting acts of professionals necessary to the final ingredients to a disaster-in-waiting that has been bubbling away within the system, often for a very lengthy period.

So why is the criminal justice system so set upon punishing errant professionals? David Marx,[35] an aircraft engineer turned lawyer, observed that over 50 years ago the US Supreme Court traced the birth of 'criminal' human error to the onset of the Industrial Revolution. Before then, both an *actus reus* (a criminal act) and a *mens rea* (a criminal intent) were necessary to achieve a criminal prosecution. But since the coming of powerful and potentially injurious machines that could, through human actions, cause significant harm, a criminal intent is no longer required to prove guilt. A driver can be prosecuted for causing a death even though the fatality arose from unwitting errors. The fact is that 'honest errors' to which we are all prone are now considered criminal in circumstances where public safety is at stake. The law has yet to adopt a systems view of human error.

The System Perspective

The following passage will serve to make the link between the person and the system perspectives. It comes, once again, from the excellent Institute for Safe Medication Practices website:

> The reasons workers drift into unsafe behaviours are often rooted in the system. Safe behavioural choices may invoke criticism, and at-risk behaviours may invoke rewards. For example, a nurse who takes longer to administer medications may be criticized, even if the additional time is attributable to safe practice habits and patient education. But a nurse who is able to handle a half-dozen of new admissions in the course of a shift may be admired, and others may follow her example, even if dangerous shortcuts have been taken. Therein lies the problem.

---

35 See newsletter of the Institute for Safe Medication Practices, 8 March, 2007.

The rewards of at-risk behaviour can become so common that perception of their risk fades or is believed to be justified.[36]

For me, though not for all, a system perspective is any accident explanation that goes beyond the local events to find contributory factors in the workplace, the organisation and the system as a whole. The essence of such a view is that frontline personnel are not so much the instigators of a bad event, rather they are the inheritors of latent conditions (or resident pathogens) that may have been accumulating for a long time previously.

The 'Swiss cheese' model[37] is only one of many such accident models. My friend Erik Hollnagel has provided a scholarly review of them in his excellent book (also published by Ashgate)[38] so I will not consider them in detail here. However, I would like to offer a somewhat different slant upon these models than the one presented by Hollnagel.

Hollnagel grouped the accident models as shown below:

- *Sequential accident models*: these included Heinrich's domino theory[39] and Svenson's accident evolution and barrier model[40] and Green's[41] 'anatomy of an accident' framework. Sequential models may involve a relatively simple linear series of causes and effects (as in the domino model) or multiple sequences of

36 Institute for Safe Medication Practices (2006). 'Our long journey towards a safety-minded just culture, part II: where we're going.' Newsletter 21 September. Available at: http://www.ismp.org/Newsletters/acutecare/articles/200609221.asp. It was also quoted in an excellent paper by Dr Larry Veltman (2007) 'Getting to Havarti: moving towards patient safety in obstetrics.' *Obstetrics and Gynecology*, 110, 5: 1146–1150.

37 Reason, J. (1997) *Managing the Risks of Organizational* Accidents. Aldershot: Ashgate.

38 Hollnagel, E. (2004) *Barriers and Accident Prevention*. Aldershot: Ashgate.

39 Heinrich, H.W. (1931) *Industrial Accident Prevention*. (New York: McGraw-Hill). There is also a more recent 5th edition: Heinrich, H.W. *et al.* (1980) New York: McGraw-Hill.

40 Svenson, O. (2001) 'Accident and incident analysis based on the accident evolution and barrier function (AEB) model.' *Cognition, Technology & Work*, 3 (1): 42–52.

41 Green, A.E. (1988) 'Human factors in industrial risk assessment – some earlier work.' In L.P. Goodstein, H.B. Andersen, and S.E. Olsen, *Tasks, Errors and Mental Models*. London: Taylor & Francis.

events as represented by a fault tree diagram (Green). They share a belief in some initial root cause, a fallacious view, according to Hollnagel.[42]

- *Epidemiological accident models*: this is the category to which 'Swiss cheese' has been assigned. The analogy here is with the spread of disease arising from interactions between several factors, some potentially evident, others latent. Although Hollnagel makes a brave attempt to describe a generic epidemiological model (p.57), I am not entirely convinced about the term 'epidemiological'. 'Swiss cheese' is primarily about how unsafe acts and latent conditions combine to breach the barriers and safeguards. I have used medical terms – resident pathogens, the body's immune system, and so on – but I still think of it as a systemic model, albeit one that retains some sequential characteristics.
- *Systemic accident models*: these, in Hollnagel's opinion, are the rightful occupants of the theoretical pantheon. Following Perrow's[43] view that accidents within complex tightly coupled systems are both normal and inevitable, Hollnagel regards accidents as emergent rather than resultant phenomena. The main distinction between emergent and resultant phenomena is that while the latter are predictable from a system's constituents, the former are not. To put it another way, while resultant phenomena can be – in theory – explained by sequential cause and effect modelling, emergent events arise out of complex dynamic networks and as such are not amenable to simple causal explanations. Notions like error and unsafe acts are banned from the analytical lexicon because of their perjorative overtones. The 'true' system approach advocates a search for unusual dependencies and common conditions as well as continuous monitoring of system variability. In the latter part of the book, Hollnagel offers stochastic resonance as a systemic model for accidents and a guide to their prevention. I cannot do justice to it here – largely because I don't entirely understand it, but that fault is mine.

---

42 And in this respect I agree with him. Each part of the 'chain' of events has its precursors: the only real root cause is the Big Bang. That being said, we have to face the fact that many organisations are using Root Cause Analysis to good effect. They are at least tracking upstream for contributory factors. Although the label may be regrettable, the activities it promotes are not.

43 Perrow, C. (1984) *Normal Accidents: Living with High Risk Technologies*. New York: Basic Books.

The groups of models outlined above are presented as a developmental or evolutionary progression going from models that by their reliance on cause-and-effect chaining miss the essence of organisational accidents to ones that capture it more accurately; from the simple-minded to the views that more truly reflect the complex and combinatorial nature of these events.

It is time to make my own position clearer. First, I believe that all of the models described by Hollnagel meet the criteria for systemic perspectives. Second, they all have – and have had – their uses. Just as there are no agreed definitions and taxonomies of error so there is no single 'right' view of accidents. In our business, the 'truth' is mostly unknowable and takes many forms. In any case, it is less important than practical utility.

To what extent does a model guide and inform accident investigators and the managers of hazardous systems who seek to prevent their occurrence? In order to meet these pragmatic requirements, I believe that a model must satisfy the following criteria:

- Does it match the knowledge, understanding and expectations of its users?
- Does it make sense and does it assist in 'sense-making'?
- Is it easily communicable? Can it be shared?
- Does it provide insights into the more covert latent conditions that contribute to accidents?
- Do these insights lead to a better interpretation of reactive outcome data and proactive process measures?
- Does the application of the model lead to more effective measures for strengthening the system's defences and improving its resilience?

Stages in the Development of the 'Swiss cheese' Model (SCM)

As indicated earlier, the SCM is but one of a large number of systemic accident models, but it is the one whose development I am best qualified to discuss, and also the model with which many readers of this book will perhaps be most familiar. It has been used in a variety of hazardous domains that are liable to suffer organisational accidents: aviation, manned space flight, rail transport, chemical process industries, oil and gas exploration and production, nuclear power plants, shipping, the US Navy and, most recently, health care. It has been around for over 20 years, and has taken a number of quite different forms. These are summarised below.

The mid-to-late-1980s version The starting point for the model[44] was the essential, benign components of any productive system: decision-makers (plant and corporate management), line management (operations, maintenance, training, and the like), preconditions (reliable equipment and a skilled and motivated workforce), productive activities (the effective integration of human and mechanical elements) and defences (safeguards against foreseeable hazards. These productive 'planes' eventually became the cheese slices – though I had no thoughts of Swiss cheese at that time.

The various human and organisational contributions to the breakdown of a complex system are mapped onto these basis productive elements as shown in Figure 5.1.[45] At that time they comprised two kinds of failure: latent failures (resident 'pathogens' within the system) and active failures (unsafe acts). The basic premise of the model was that organisational accidents have their primary origins in the fallible decisions made by designers, builders and top-level management. These are then transmitted via the intervening productive elements – line management deficiencies, the psychological precursors of unsafe acts, the unsafe acts themselves – to the point where these upstream influences combine with local triggers and defensive weaknesses to breach the barriers and safeguards.

Although some people still seem to prefer this view, I believe it has two serious limitations. Despite my original protestations[46] that this was not a way of shifting blame from the 'sharp end' to the board room, it still looks very much like it. Strategic decisions

---

44 Reason, J. (1990) *Human Error*. New York: Cambridge University Press.

45 This early version of the model was greatly influenced by my friend, John Wreathall, a nuclear engineer by training but now a human and organisational factors consultant. I feel I have never sufficiently acknowledged his huge contribution to the early version of the model. I hope I have done him justice here.

46 See Reason, J. (1990), p. 203.

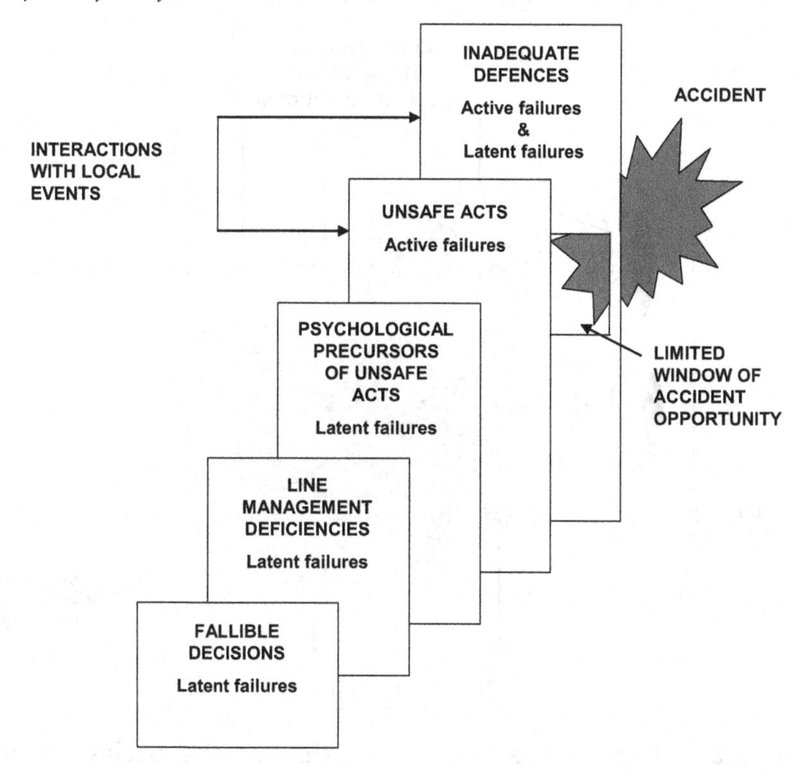

Figure 5.1 First version of the Swiss cheese model (though it hadn't yet taken on its Emmenthale appearance. Various human contributions to the breakdown of complex systems are mapped on to the basic elements of production)

may be fallible but they need not be. The second problem is its simple cause and effect linearity that inevitably leads, as Hollnagel has pointed out, to a search for 'root causes' (see footnote 41 on page 94 for an earlier discussion of this problem). Back in the late 1980s, I added another picture to the development of the earliest version (see Figure 5.2) that has a distinctly 'cheesy' flavour to it, though that was not intended at the time.[47]

I am embarrassed to admit that only I have only recently realised from whence the 'Swiss cheese' label might have sprung.

---

47 I didn't invent the label 'Swiss cheese model' though I'm eternally grateful to the person or people who did. I have two 'suspects' for the role of inventor: Dr Rob Lee, then Director of the Bureau of Air Safety Investigation in Canberra, and Captain Dan Maurino, the human factors specialist at ICAO in Montreal.

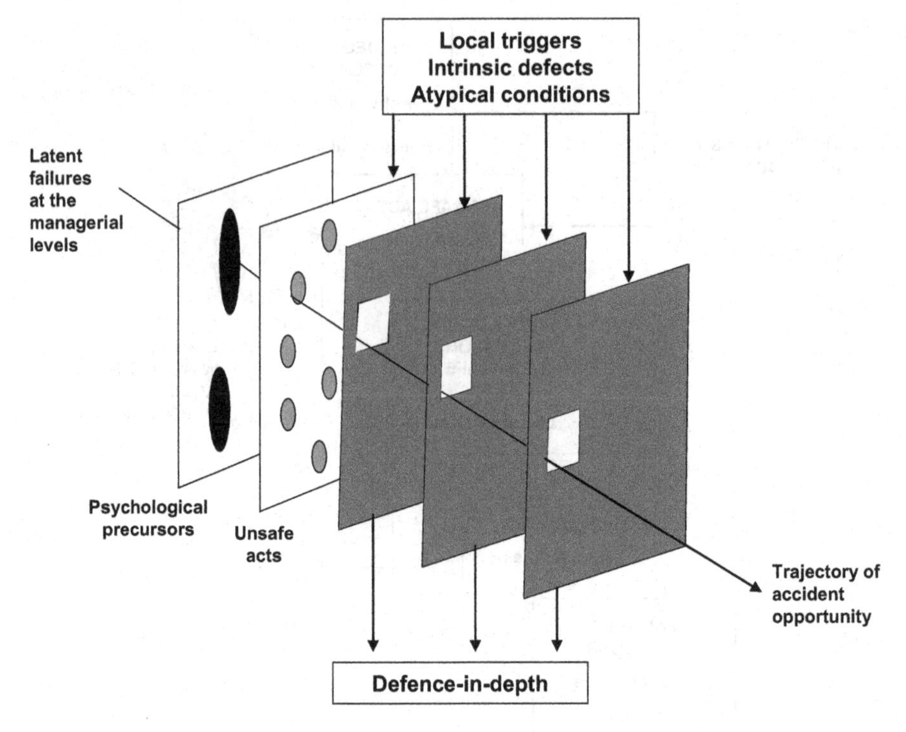

Figure 5.2 Part of the earliest version of the Swiss cheese model. The diagram shows a trajectory of accident opportunity penetrating several defensive layers, and begins to have Emmenthale-ish features

I think it has to be Figure 5.2, although it only includes the last few slices. It was there that I adopted the pictorial convention of showing the precursors and unsafe acts as holes through which an accident trajectory could pass.

Early 1990s version A later variant (see Figure 5.3) assumed that a variety of organisational factors could seed latent pathogens into the system. These included management decisions, core organisational processes – designing, building, maintaining, scheduling, budgeting, and the like – along with the corporate safety culture. The significant thing about culture is that it can affect all parts of the system for good or ill. There were two ways in which the consequences of these upstream factors could impact adversely upon the defences. There was an active failure pathway in which error- and violation-producing conditions

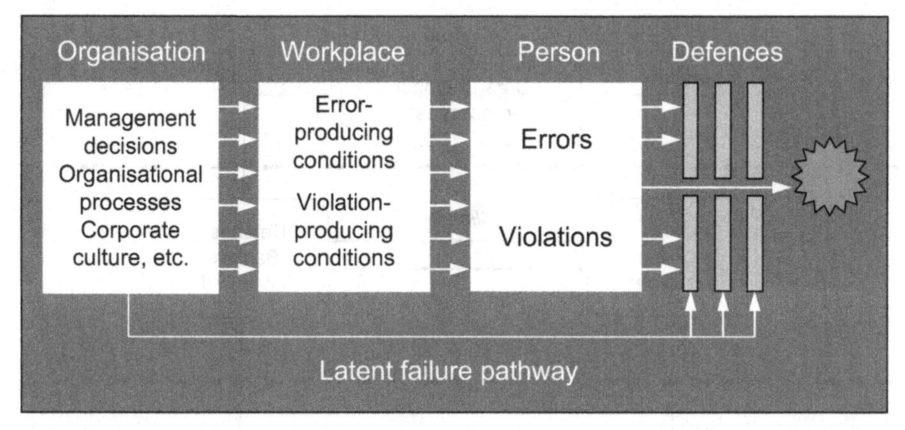

Figure 5.3 An early 1990s variant of the Swiss cheese model

in the workplace could, at the individual or team level, create unsafe acts. A very large number of unsafe acts are likely to be committed, but only very few of them will find chinks in the systems defences.

There was also a latent failure pathway that transmitted pathogens to the defences directly. Unsafe acts at the sharp end are not essential – though common – for creating defensive gaps and weaknesses, as is evident from the King's Cross Underground fire, for example.

A mid-1990s variant Aside from making an appearance in a 1995 book[48] and forming the basis of RAIT (Railway Accident Investigation Tool),[49] this rather more elaborate version of the model has not had much of an airing, though I still think it has some merits as an incident and accident analysis guide. It is summarised in Figure 5.4.

The principal innovations were, firstly, the identification of two distinct failure pathways, the human error pathway and the defensive failure pathway; and, secondly, the restriction of the term *latent failure* to weaknesses or absences in the defences, barriers and safeguards. This scheme also required a clear separation of defensive functions from organisational processes. Unlike

48 Maurino, D., Reason, J., Johnston, N., and Lee, R. (1995) *Beyond Aviation Human Factors*. Aldershot: Ashgate Publishing Ltd.

49 Reason, J., Free, R., Havard, S., Benson, M., and van Oijen, P. (1994) *Railway Accident Investigation Tool (RAIT): A Step-by-Step Guide for New Users*. Manchester: Department of Psychology, University of Manchester.

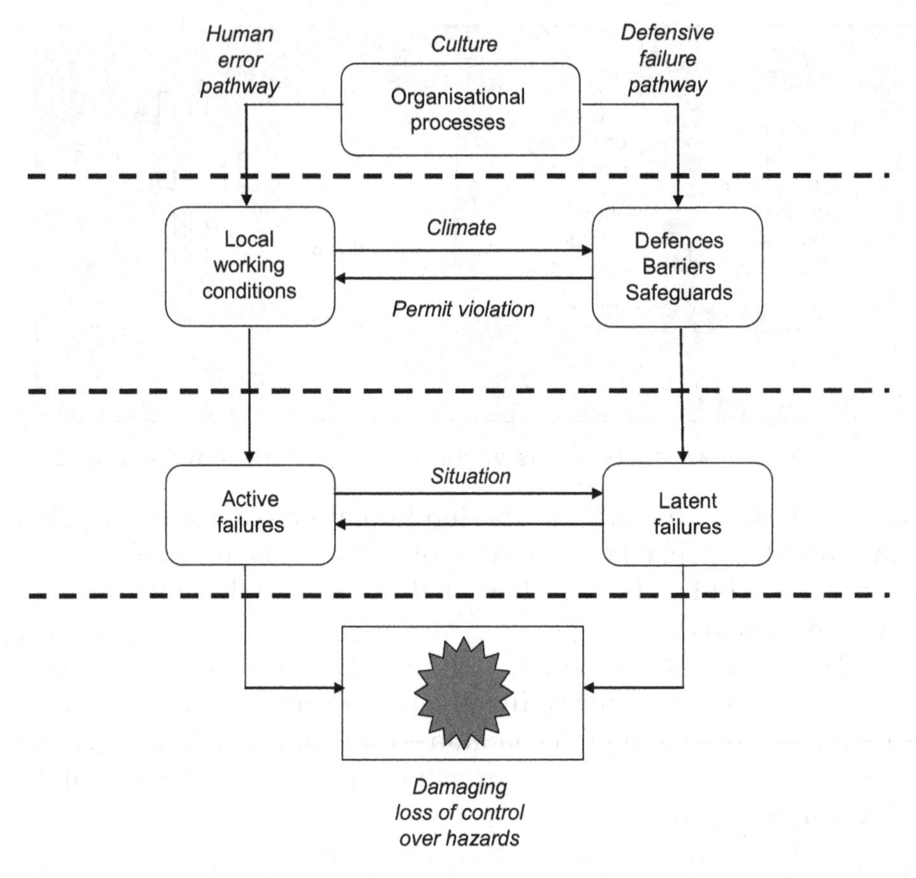

Figure 5.4 A mid-1990s variation with two interacting pathways to an accident

previous representations of the model, the causal sequence runs from top to bottom (rather than from left to right). The accident sequence is divided into four levels: culture, climate, situation and the event itself.

Both the human error and the defensive failure pathways have their origins (within the system, at least) in the organisational processes. These include goal-setting, policy-making, organising, forecasting, planning, scheduling, managing, financing, allocating resources, communicating, designing and specifying. All of these processes are likely to reflect cultural influences, and each them can contribute to a breakdown in human performance or a defensive failure.

Whereas *culture* emanates from the 'strategic apex' of the organisation, *climate* relates to specific workplaces and to their local management, resources and workforce. Climate is shaped both by upstream cultural factors and the local circumstances. The latter are sub-divided into two interacting sets of factors: local working conditions and the defences existing within that particular location.

Unlike culture, local climates can change quite rapidly. Each flight crew in a commercial aircraft will create a different climate. Similarly the arrival of a new master aboard a merchant ship (even a different watchkeeper) or a new toolpusher on a drilling rig will bring about a change of climate. Local managers play a large part in determining the psychological working conditions and also in influencing how effectively the available defences are employed.

The Current Version

The current model (1997) involves a succession of defensive layers separating potential losses from the local hazards. As can be seen from Figure 5.5 this version finally acknowledges the Swiss cheese label. Each 'slice' – like Emmenthale – has holes in it; but unlike cheese the gaps are in continuous motion, moving from place to place, opening and shutting. Only when a series of holes 'line up' can an accident trajectory pass through the defences to cause harm to people, assets and the environment. The holes arise from unsafe acts (usually short-lived windows of opportunity) and latent conditions. The latter occur because the designers, builders, managers and operators cannot foresee all possible accident scenarios. They are much more long-lasting than the gaps due to active failures and are present before an adverse event occurs.

There were two important changes. First, the defensive layers were not specified. They included a variety of barriers and safeguards – physical protection, engineered safety features, administrative controls (regulations, rules and procedures), personal protective equipment and the frontline operators themselves: pilots, drivers, watchkeepers and the like. They often constituted the last line of defence. The second change was the use of the term 'latent conditions'. Conditions are not causes, as such, but they are necessary for the causal agents to have their

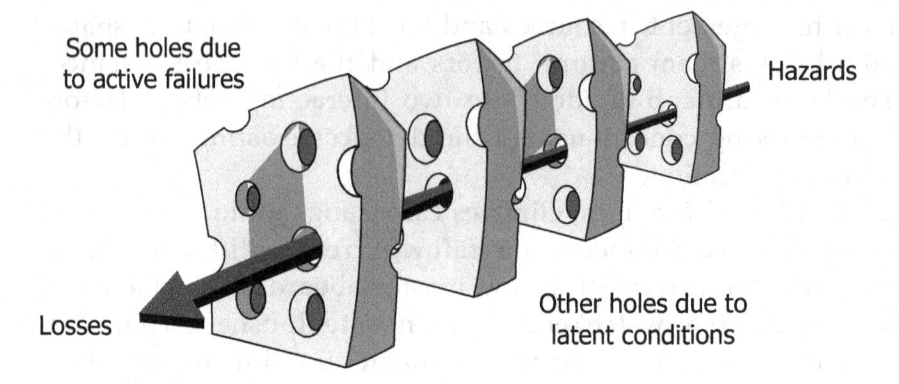

Figure 5.5 The latest version of the Swiss cheese model

effects. I will discuss this distinction further and elaborate upon the current model in later chapters.

Person and System Models: Getting the Balance Right

We have already discussed the weaknesses of the person model at some length, all of which relate to the 'human-as-hazard' perspective. The 'human-as-hero' view is quite another matter and will be considered extensively in the next part of this book.

Although the system models seem, on the face of it, to be far more appropriate ways of considering accident causation, both in terms of understanding the contributing factors and in their remedial implications, they too have their limitations when taken to extremes. This was first brought home to me by the brilliant essays of Dr Atul Gawande, a general surgeon at a large Boston hospital and a staff writer on science and medicine for the *New Yorker*.[50]

In an essay entitled 'When doctors make mistakes',[51] Dr Gawande recounts the many successes of American anaesthesiologists in

50 Atul Gawande's articles for the New Yorker are collected in two wonderful books: *Complications: A Surgeon's Notes on an Imperfect Science* (New York: Metropolitan Books, 2002) and *Better: A Surgeon's Notes on Performance* (New York: Profile Books, 2007).

51 Gawande (2002), pp. 47–74.

reducing adverse events through various systemic measures involving the redesign and standardisation of anaesthetic machines, the use of pulse oximeters, carbon dioxide monitors and improved training measures employing high-fidelity anaesthesia simulators. In ten years, the death rate dropped to one-twentieth of what it had been. He concludes his discussion with the following very telling passage:

> But there are distinct limitations to the industrial cure, however necessary its emphasis on systems and structures. It would be deadly for us, the individual actors, to give up our belief in human perfectibility. The statistics may say that someday I will sever someone's main bile duct [a recurrent error in laparoscopic cholecystectomy procedures], but each time I go into a gallbladder operation I believe that with enough will and effort I can beat the odds. This isn't just professional vanity. It's a necessary part of good medicine, even in superbly 'optimized' systems. Operations like the lap chole have taught me how easily error can occur, but they've also showed me something else: effort does matter; diligence and attention to the minutest details can save you.[52]

This brings us to the nub of the problem with regard to an excessive reliance on system measures. People on the frontline of health care or any other hazardous enterprise generally have little opportunity to bring about rapid system improvements, or any kind of global change, but they can resolve to go the extra mile. Health care, in particular, has a one-to-one or few-to-one delivery. It's a hands-on and very personal business. Personal qualities do matter. To think otherwise is to fall prey to 'learned helplessness' – saying to oneself 'What can I do? It's the system'.

I also believe that we can train frontline people in the mental skills that will make them more 'error wise'. That is, to help them to 'read' situations so that they can identify circumstances having high error potential, and act accordingly. This is a perception of 'sharp-end' individuals that accentuates the positive so that we can exploit their 'human-as-hero' potential. This is what we mean by individual mindfulness: being aware of the hazards and having contingencies in place to deal with them; being respectful of the dangers and possessing a 'feral vigilance' in their presence. We will discuss these issues in detail in the final part of this book.

52 Ibid, p. 73.

PART III
Accidents

Chapter 6

Error Traps and Recurrent Accidents

Let me begin by clarifying what exactly is meant by the two terms that make up the title of this chapter. If you were to google 'error traps' you would find that the vast majority of the hits (about 1,700,000 in total at my last look) related to computing, where the term is used to describe ways of trapping programming bugs. This is not my meaning here. For me, and other human factors people, the term denotes situations in which the same kinds of errors are provoked in different people. Here, for the most part, we will be discussing error-prone situations rather than error-prone people or software devices for catching coding errors.

Similarly, if you were to google 'recurrent accidents' you would find the term used in two different senses: accident-liable individuals and situations that produce repeated accidents. My main interest here is in the latter; but I will also look briefly at accident-proneness as it relates to certain individuals.

Accident-proneness: A Quick Survey

The idea that certain people, by dint of enduring personality characteristics, are more liable to misfortune goes back at least to the Book of Job and has wide popular appeal. Most of us think we know such people. Yet the notion of accident-proneness has a vexed scientific history that does not, on balance, support the idea of some relatively fixed traits that render certain individuals more likely to suffer accidents.

What is less contestable is the idea of unequal accident liability in particular situations. When the accidents of a group of individuals having equal exposure to the risks are counted and compared with the chance expectation (as determined by the Poisson distribution – see Figure 6.1), it is found that a few

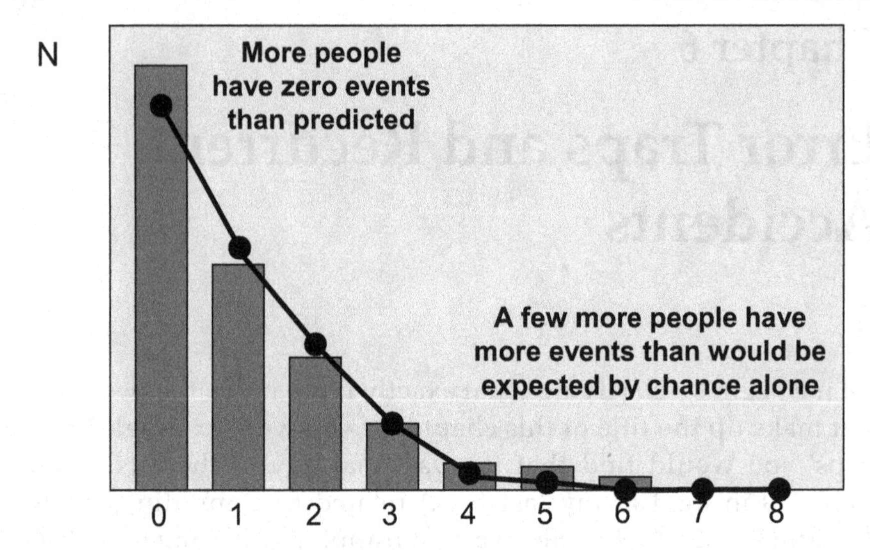

Figure 6.1 Showing the unequal liability to accidents in a group exposed to the same hazards over the same time period. The curve represents the chance prediction of event likelihood (Poisson distribution). The columns represent the actual number of accidents sustained by the group

individuals have more than their 'fair share' of adverse events. These uneven distributions of accidents have been found in a number of different circumstances – ranging from women working in a chocolate factory[1] in the First World War to contemporary Boeing 747 pilots in a large national airline.[2] But 'accident-proneness' appears to be a club with a shifting membership.

Attempts to link accident liability to particular personality characteristics have been largely unsuccessful, although a number of studies have found weak links between repeated accidents and unstable or sociopathic extraversion (self-centred, over-confident, aggressive, irresponsible, impulsive and antagonistic to authority)

---

1 Greenwood, M., and Woods, H.M. (1919) *A Report on the Incidence of Industrial Accidents With Special Reference to Multiple Accidents*. Industrial Fatigue Research Board Report No. 4. London: Her Majesty's Stationery Office.

2 Savage, J. (1997) 'FlightData recording in British Airways.' *Flight Deck*, 23: 26–34.

and also anxious (unstable) introverts (indecisive, tense, depressed, difficulties with concentration, easily intimidated, feelings of inadequacy). It is clear, however, that the link between personality and accident liability is likely to be indirect and influenced by a host of other factors.[3]

My preferred way of interpreting unequal accident or event liability is to look for clusters. Sometimes the accident clusters are associated with particular conditions in the workplace. For example, some pilots may have difficulties with particular runways on specific airfields, or be prone to level busts (overshooting the altitude directed by the air traffic controller), or have a tendency to make heavy landings. These situational or task-related clusters suggest the need for focused retraining.

Other clusters may not be associated with the task or the environment, rather they are bunched in a given time period. This suggests that the individual is going through a bad patch of one kind or another, and is suffering from what Morris Schulzinger called the 'accident syndrome'[4] for which counselling might be appropriate. Of course, there might also be people whose accidents are not associated with situations or bunched over time. I have suggested, tongue-in-cheek, that these individuals should be promoted to management, or to some other position in which they can do little direct harm.

Farmer, who pioneered the notion of dispositionally determined accident-proneness in the 1920s, concluded after a lifetime of research that 'it is probable that no unique accident prone personality exists, and that progress will be made only by attempting to relate personal characteristics to specific types of human error.'[5]

In the 1980s, research groups in Oxford and Manchester used self-report questionnaires to examine individual differences in proneness to absent-minded action slips and memory lapses.[6] In

---

3 Lawton, R., and Parker, D. (1998) 'Individual differences in accident liability: A review and integrative approach.' *Human Factors,* 40 (4): 655–671.

4 Schulzinger, M. (1956) *The Accident Syndrome: The Genesis of Accidental Injury*. Springfield IL: Charles C Thomas.

5 Farmer, E. (1984) 'Personality factors in aviation.' *International Journal of Aviation Safety,* 2: 175–179. Cited by Lawton and Parker (1998) ibid.

6 This research is discussed in detail in: Reason, J. (1993) 'Self-report questionnaires in cognitive psychology: Have they delivered the goods?' In A. Baddeley and L. Weiskrantz (eds) *Attention: Selection, Awareness, and Control.* Oxford: Clarendon Press (pp. 406–423).

general, these questionnaires described a wide range of minor slips and lapses, and asked people to indicate along some ordinal scale how often a particular kind of error crops up in their daily lives.

Although the specific forms of these questionnaires differed from one research group to another, as did the types of people to whom they were given, the results from the various studies showed a surprising degree of agreement – surprising, that is, for this kind of psychological research. The main findings are summarised below:

- There is strong evidence to indicate that people do, in fact, differ widely in their proneness to absent-minded errors. That this is not simply a question of how they would like to present themselves is shown by their spouses' general agreement with their self-assessments. Moreover, the indications are that this characteristic liability to error is a fairly enduring feature of the individual, at least over a period of 16 months or more.
- Liability to minor cognitive failures spans a wide range of mental activities and does not appear to be specific to any one domain such as memory, action control and so forth. Thus, individuals who acknowledge that they have more than their fair share of memory lapses also report making a relatively large number of errors resulting from failures of attention and recognition, and conversely. It would seem, therefore, that susceptibility is determined by the characteristic way in which some universal mental control process – the limited attentional resource perhaps – is deployed. This appears to operate relatively independently of the particular parts of the cognitive system in which it could show itself.
- The Oxford group produced tentative evidence to indicate that error-prone people are more vulnerable to stress. That is, they are likely to show minor psychiatric symptoms – fine cracks in the mental fabric – sooner than those who are less liable to minor slips and lapses. This does not appear to have anything to do with the errors themselves, rather that a high rate of minor cognitive failures might reflect a chronic inefficiency in the deployment of attention, which is brought into greater prominence by stressful conditions.

To date, however, there has been no conclusive evidence to suggest that these marked individual differences in absent-minded slips and lapses are directly related to accident liability. The highest correlation was with age. Paradoxically, self-reported liability to absent-mindedness declines markedly with advancing years – perhaps because the more mature of us acknowledge our 'senior moments' and rely heavily on memory aids (diaries, lists, knotted handkerchiefs, and the like) in order to limit their occurrence. Or maybe it's because the older you get, the more likely you are to forget your absent-minded episodes. But I don't really think so: my absent-minded embarrassments remain firmly etched in the mind – if only I could forget them.

Everyday Error Traps

Our house has on its four floors eight washbasins and sinks. Seven of them are plumbed in correctly according to the European way: the cold tap on the right and the hot tap on the left. But in my bathroom, the one I use most frequently, the order is reversed. I don't have a problem when I'm there. The errors occur when I use any of the remaining tap configurations. I cannot count the number of times I have sought a glass of cold drinking water from the hot tap. This is a case of negative transfer, one of the most potent of all error-producing factors.

Here is another example of negative transfer. Most European cars have the direction indicator and light controls to the left of the steering wheel with the windscreen wiper switches on the right. On Japanese cars these locations tend to be reversed. How often, when moving from one type of car to another, do you find yourself attempting to signal a turn by activating the windscreen wipers, and conversely?

Change, either in a set of routine actions or in the location of frequently used objects, is another powerful error-producer. Wanting to lose weight, I resolved to drink only one glass of wine with dinner, but my hand automatically reaches out and pours a second (or third) glass.

A few years ago, my wife and I decided to carry out an experiment to test the strength of well-established habits. To this end, we swapped the location of the cutlery drawer with the one on its immediate left. Three months later we were still trying to retrieve knives and forks from the wrong drawer – although not on every occasion. These absent-minded slips are almost always accompanied by 'attentional capture' at the point when some conscious intervention is necessary to redirect the habitual actions.

A recurrent source of wrong-dose errors in hospitals is the wide variety of infusion pumps and syringe drivers that are in use within the same institution. Some widely used devices are calibrated in mm per hour; others are calibrated in mm per day. Busy health carers working in distracting surroundings are quite literally 'set up' to deliver the wrong dose of medication, sometimes with tragic consequences.

Picture a simple desktop photocopier; this is an almost irresistible error trap. Suppose you wanted to copy a loose-leaf document of some dozen or so pages – what is the most likely error? It is departing the scene with the completed copy and leaving the last page of the original on the platen under the closed lid. Four task-related error-provoking factors make this omission nearly inevitable:

1. *Premature exit*: Steps near the end of a routine task are more prone to omission because our minds are on the next job.
2. *Lack of cueing*: Removing previous original sheets was cued by the need to place the next one on the platen. Such prompting is not available for the last sheet.
3. *Goal achieved before the task is complete*: The purpose of the task is to make a copy. The duplicated sheets are clearly visible on the out-tray.
4. *Out of sight, out of mind*: The last sheet of the original is concealed under the lid.

Another kind of error trap is created by priming. In the example below, the trap is created by a sequence of questions yielding (for the most part) same-sound answers (see also Chapter 3):

- What kind of tree grows from an acorn? [oak]
- What do you call a funny story? [joke]
- What rises from a bonfire? [smoke]

- What's another name for a cape? [cloak]
- What kind of noise does a frog make [croak]
- What do you call the white of an egg? [ ]

When the questions are asked in manner that demands a rapid answer, the large majority of respondents will reply to the last question with 'yolk' – an incorrect response. Other factors also conspire to induce this error. The correct answer 'albumin' is a relatively infrequent word in most people's lexicon. Another right answer 'white' is already present in the question and it is unusual to respond with a word embedded in the question. Finally, the likelihood of replying with 'yolk' is dependent on the number of prior 'oak-sound' primes. The more there are, the greater the probability of making an error.

Recurrent Accident Patterns

Those who spend time examining accident reports will know the 'not again!' reaction. Far from being random, accidents in many different domains have a way of falling into recurrent patterns. If the same types of accident keep happening, then many different people are involved. The critical causal features must therefore implicate the local factors: the task and the working environment, together with the nature of the system as a whole. Consider the following examples drawn from a variety of technologies.

Railways

My interest in error traps and recurrent accident patterns was originally sparked off by the following passage from LTC Rolt's marvellous book *Red for Danger*:

> The mistake which has probably caused more serious accidents in the last sixty years than any other is that of a signalman forgetting, especially at night, that he has a train standing near his box. It may be a train that has halted in obedience to his danger signals, or it may be one that he has 'wrong roaded', that is to say transferred to the wrong running line to wait there until a fast following train has passed.[7]

7 Rolt, L.T.C. (1978) *Red for Danger*. London: Pan Books (p. 194).

A terrible example of this type of error occurred at Quintinshill, just north of Gretna Green, on the Carlisle–Glasgow line in 1915. This was Britain's worst railway disaster. Two hundred and twenty-six people were killed, most of them soldiers en route to the Western Front.

The curious feature of this accident was that the train which the signalman forgot was the slow passenger train (now standing by the signal box) that he had hitched a ride on from Gretna Green in order to start his shift. The signalman he was replacing had also failed to place reminder collars over his signal levers, which was normal practice when a line was blocked by a standing train. The signalman who had just arrived in the signal box accepted the up (southbound) troop train. Two minutes later it smashed into the parked passenger train with great violence and the wooden coaches of the troop train caught fire. Seconds later the northbound Scottish express crashed into the wreckage.

I mentioned signals passed at danger (SPADs) in Chapter 3 and will revisit them here. A SPAD is logged when a train has run beyond its allocated signal block without authority. Typically, the block is indicated by a line side signal at red. In 1990, British Rail (BR) researchers[8] in Derby carried out a detailed analysis of SPADs occurring on the British Rail network in the period 1986 to 1989. These were the findings:

- Far more signals were repeatedly passed at danger than would be expected by chance ($p<.000001$).
- Ninety-three per cent of BR's 36,000 signals were SPAD-free.
- One per cent of signals attracted 30 per cent of the SPADs.
- 0.3 per cent of the signals attracted 15 per cent of the SPADs.

The conclusion is clear: certain signals are more 'SPAD-prone' than others; some very much more so. Certain features of signals make them SPAD-provoking than others. These include: location, mounting, visibility, number of aspects and arrest rates. In short, certain signals constitute recurrent error traps. Since 1913, there have been at least 16 accidents worldwide resulting from the unauthorised passing of a red light. These include the Southall

8 BR Research (1990) *An Analysis of Signals-passed-at-danger, 1986–1989*. Derby: BR Research Report.

rail crash in 1997 and the Ladbroke Grove rail crash that occurred two years later in a similar locality.

Aviation Examples

Controlled flight into terrain (CFIT) occurs when the flight crew loses situational awareness in proximity to the ground. CFIT accidents account for about 74 per cent of fatal accidents worldwide in commercial aviation. A study by the Flight Safety Foundation[9] discovered the following recurrent features:

- More than half of all CFIT accidents occur during step-down instrument approaches involving intermittent descents at relatively steep angles, or during approaches at abnormally shallow angles – less than one degree.
- About half of all CFIT accidents involve freight, charter or positioning flights.
- About half of all CFIT accidents involve less than three per cent of the world's total fleet. This three per cent are not equipped with ground proximity warning systems (GPWS). It should also be noted that less than 30 per cent of the corporate and business fleet is equipped with GPWS.
- Inadequate charts together with significant differences between government- and commercially produced charts were regarded as a significant causal factor in many CFIT accidents.

Nuclear Power Generation

Surveys carried out by the Institute of Nuclear Power Operations in the US[10] and the Centre for Research in the Electrical Power Industries in Japan[11] revealed that something in excess of 70 per cent of all human factors problems in nuclear power plants were maintenance-related. Of these, the majority involved the

9 See BASI (1995) *Israel Aircraft Industries Westwind 1124 VH-AJS Alice Springs Northern Territory, 27 April 1995.* Investigation Report No. 9501246. Canberra, ACT: Bureau of Air Safety Investigation.

10 INPO (1984) *An Analysis of Root Causes in 1983 Significant Event Reports.* INPO 84-027. Atlanta, GA: Institute of Nuclear Power Operations. INPO (1985) *An Analysis of Root Causes in 1983 and 1984 Significant Event Reports.* INPO 85-027. Atlanta, GA: Institute of Nuclear Power Operations.

11 See Takano, K. (1996) Personal communication.

omission of necessary steps during installation or re-assembly. Similar recurrent patterns of error were also observed in aircraft maintenance.[12] In an analysis of critical incident reports from experienced aircraft maintainers, Hobbs[13] also found omissions to be the single largest category of error. The most common local factors leading to these and other errors were inadequate tools and equipment (e.g., broken stands and faulty electrical devices), perceived pressure or haste, and environmental conditions such as bad weather, darkness and slippery work surfaces. In a study of ground damage incidents to aircraft, Wenner and Drury[14] found similar error-affording factors: poor equipment and not using the correct number of personnel to carry out the job.

Another type of recurrent event has been the loss of residual heat removal (RHR) at mid-loop in a pressurised water reactor (PWR). During maintenance and refuelling operations, a PWR may have the primary system water level reduced to the point where the hot leg is only partly full. Under these circumstances, the decay heat removal path is through the RHR system, with heated water from the core flowing to the partly full hot leg and thence down to the RHR pumps, heat exchangers and return path to the cold leg. At such times, a small error in adjusting the desired water level can lead to drainage below the hot leg, resulting in an interruption of the heat transport to the RHR heat exchanger system. There were 20 such events in the time period 1980–96. That is more than one per year. Although the events were widely publicised within the US nuclear industry, the same type of bad scenario continued to occur, even though the corrective actions were well known.[15]

In 1993, I was asked by the US Nuclear Regulatory Commission to review 21 human factors analyses of significant events occurring

12 Reason, J. (1995) *Comprehensive Error Management in Aircraft Engineering: A Manager's Guide*. London Heathrow: British Airways Engineering.

13 Hobbs, A. (1997) *Human Factors in Airline Maintenance: A Study of Incident Reports*. Canberra, ACT: Bureau of Air Safety Investigation.

14 Wenner, C., and Drury, C.G. (1996) 'Active and latent failures in aircraft ground damage incidents.' *Proceedings of the Human Factors and Ergonomics Society 40th Annual Meeting* (pp. 796–801). San Diego, CA: Human Factors and Ergonomics Society.

15 CSNI Technical Opinion Papers. No. 3. (2003) *Recurrent Events*. Paris: Nuclear Energy Organization for Economic Co-operation and Development.

in US commercial nuclear power plants.[16] These analyses indicated that particular organisational weaknesses are likely to be made more apparent by certain plant states, and that these, in turn, shape the kinds of unsafe acts likely to be committed by plant operators. Of particular interest were the dependencies that are revealed during low power and shut down (LP & S) events (i.e., when the reactor is sub-critical or at less than 10–15 per cent of full power). Although such states are atypical – they represent only a small proportion of the time that the reactor is operating at full power (or close to it) – they are not abnormal. Indeed they are a necessary and planned part of nuclear power plant operation, as during maintenance or refuelling.

- The plant or technology is operated 'outside the rules'.
- The physical regime is not well understood.
- And the operators refuse to believe the available evidence indicating the true nature of the event.

These three factors were echoed in many of the less successful recoveries. Crews were forced to 'wing it' because the procedures were either inappropriate or unavailable. As a result they were forced to improvise in conditions which they did not fully understand. And like others in similar circumstances they were prey to confirmation bias, or fixation errors that prevented them from recognising evidence at odds with their current mindsets. Thus, it was found that mistakes (planning and problem-solving failures) were the dominant error mode in the LP & S events. In addition, errors of commission (performance of an action that causes the plant to be in a less safe state) were far more common in the LP & S states than were errors of omission (failures to perform a necessary action).

These findings make it possible to outline a sequence of qualitative dependencies linking the organisational decision makers, via the plant state, to operator errors. Although LP & S states are anticipated, the plant managers were relatively ill-prepared for the associated events. As a consequence, the

16 NUREG/CR-6093 (1994) *An Analysis of Operational Experience During Low Power and Shutdown and a Plan for Addressing Human Reliability Assessment Issues*. Rockville, MD: US Nuclear Regulatory Commission.

operators tended to be lacking in both state-specific training and relevant emergency procedures. Furthermore, these outages were often poorly planned, despite the presence of a much larger staff working within the power plant at these times. Crews were forced to improvise recovery strategies, and these attempts were often mistaken leading to a preponderance of errors of commission. Moreover, LP & S states and their associated events demanded a higher degree of direct action on the part of the operators with greater opportunities for error – particularly errors of commission that are relatively infrequent in full power events where engineered safety features tend to dominate the recovery process.[17]

Marine Transport

In a review of 1991–1995 marine accidents in Australian waters, the Marine Incident Investigation Unit[18] identified recurring causal patterns for both groundings and collisions – the two most common accident types, accounting for 60 per cent of all incidents. Of the groundings, 76 per cent occurred either in the inner route of the Great Barrier Reef (Torres Straits) or close to, or within, port limits. Thirty-three per cent of the incidents involved ships where the pilot was on the bridge directing navigation (i.e., pilot-assisted accidents). On those occasions when a pilot was not aboard, 36 per cent occurred between midnight and 0400 hours. Among the collisions, 83 per cent were between large trading vessels and small fishing boats or yachts. The collisions were facilitated by the failure to keep a proper lookout by all the vessels concerned and, in many cases, by a lack of knowledge of the International Collision Regulations by those operating the smaller vessels.

Health Care

There are few hazardous domains that bear out the tragic truth of Murphy's Law more than health care: namely, if something can

---

17 Reason, J. (1995) 'A systems approach to organizational error.' *Ergonomics*, 38: 1708–1721.

18 MIIU (1996) *Marine Incident Investigation Unit 1991–1995*. Canberra, ACT: Department of Transport and Regional Development.

go wrong, it will go wrong. Three derivatives of the law account for a large proportion of health care's recurrent fatal events:

1. If there are two ways of administering a drug, then there will be occasions when it is delivered by the wrong route in error.
2. If two ducts appear visually very similar during laparoscopic surgery, then there will be times when the wrong duct is cut in error.
3. Since the human body is largely symmetrical, then there will be times when the wrong organ or the wrong limb will be removed in error.

There are many other possible derivatives of Murphy's Law, but these three have been selected because they occur repeatedly and elicit extremely damaging errors. I will examine each of them in turn.

Wrong Route

Again, there are many possible ways of wrongly administering a drug, but the error that has occurred many times in many countries involves vincristine, a neurotoxic medicine used in the treatment of leukaemias and lymphomas. It must be administered intravenously, but on over 50 occasions it has been given intrathecally (into the spine). The error results in a fatal outcome in nearly 90 per cent of cases. Those patients that survive usually suffer permanent paralysis due to damage to the nervous system. Both adults and children are at risk.

The most frequent error is to mistake vincristine with an injection to be given intrathecally (methotrexate, for example). As a result it is administered into the cerebrospinal fluid instead of, or in addition to, other medicines. These incidents have a number of common contributing factors:

- *Same time* – prescription of intravenous vincristine in treatment protocols that require drugs to be given intrathecally on the same day and frequently at the same time. The dangers of the intrathecal administration are well known to hospitals and protocols have been developed that require intravenous and intrathecal drugs to be given on different days and to be stored in different packages. However, these defences are sometimes defeated by what I have termed 'the lethal convergence of benevolence'. Nursing staff on day wards know that some outpatients, particularly adolescent boys, are poor attenders. It is hard enough to get them to come for treatment at the best of times, but procedures that require them to attend twice are seen as impractical; so nurses and pharmacists 'collude' to have both intravenous and intrathecal drugs available for same-day administration, and even in the same package.[19]

- *Same place* – transport, storage and administration of intravenous vincristine in the same location as medicines required to be administered intrathecally.
- *Inadequate checking* – medicine labels are not always checked against treatment orders when selecting medicines from storage locations (the ward refrigerator) immediately prior to administration.
- *Insufficient knowledge* – junior staff lacking in knowledge and experience are often delegated to manage chemotherapy. The dangers of intrathecal vincristine are not always understood by senior house officers and specialist registrars. They end up as the last line of defence in a system with breached or lacking safeguards. It was one well-publicised and well-investigated case of intrathecal vincristine that prompted me to start developing mental skills training that make people more aware of high-risk situations (see footnote 19).

Wrong Duct

Cholecystectomy involves the removal of the gall bladder, a floppy sac of bile lying under the liver, frequently because gall stones have formed in the sac. Twelve years ago, the surgeon would have made a six-inch incision in the abdomen that would require the patient to have a week's stay in hospital to heal. Now, with the advent of laparoscopic surgery, all that is required are

19 Toft, B. (2001) *External Inquiry into the Adverse Incident that Occurred at Queen's Medical Centre, Nottingham, 4th January 2001*. London: Department of Health. See also Reason, J. (2004) 'Beyond the organizational accident: the need for "error wisdom" on the frontline.' *Quality and Safety in Health Care*, 13: ii28–ii33.

a few tiny incisions that enable the surgeon to insert a miniature camera and the required instruments. The procedure is guided by a TV screen. This operation, often done as day surgery, is termed laparoscopic cholecystectomy or 'lap chole'.

But there is a recurrent problem. Even experienced surgeons cut the main bile duct about once in every 200 lap chole procedures. Let Atul Gawande, a surgeon, describe the situation. 'There's one looming danger, though: the stalk of the gall bladder is a branch off the liver's only conduit for sending bile to the intestines for the digestion of fats. And if you accidentally injure this main bile duct, the bile backs up and starts to destroy the liver. Between 10 and 20 per cent of patients to whom this happens will die. Those that survive often have permanent liver damage and can go to require liver transplantation.'[20]

Dr Lawrence Wray and his colleagues analysed 252 laparoscopic bile duct injuries. The primary cause of error in 97 per cent of cases was a visual perception illusion of a very compelling nature. The illusion was described as follows: 'one could see [from the videotapes] that the CBD [common bile duct] became prominent as traction was placed on the gallbladder at the start of the dissection. The cystic duct was partially or completely hidden from view ... The resulting arrangement made the common duct appear as if it continued directly into the base of the gall bladder. In other words, the anatomic relationship between the CBD and the gallbladder mimicked the surgeon's mental model of the relationship between the cystic duct and the gallbladder.'[21]

Bile duct injuries did occur in the days before the lap chole technique, but they were relatively uncommon. Something about the laparoscopic environment predisposes surgeons to experience this compelling misperception. Many factors could contribute – limitations in perspective and the loss of stereoscopy – but Way and his colleague regard the loss of haptic perception (loss of direct feel) as the most likely candidate. If one merely brushes up against something, the experience is passive touch; active touch

---

20 Gawande, A. (2002) *Complications [A Surgeon's Notes on an Imperfect Science]*. New York: Metropolitan Books, (p. 71).

21 Way, L.W. *et al* (2003) 'Causes and prevention of laparoscopic bile duct injuries.' *Annals of Surgery*, 237: 460–469; p. 464.

occurs when one manually explores an object. Thus, even though it is hidden by connective tissue, the surface of the gallbladder can be felt and visualised.

Visual perception involves making assumptions derived from heuristic processes. Heuristics are essential for simplifying a complex world. They work well most of the time, but they can also lead to error. The Kanizsa Triangle shown in Figure 6.2 demonstrates this. Most people see an illusory white triangle in the centre of the picture. The brain automatically assumes that the shape of the three black objects arises from their being occluded by a white triangle. The process is automatic and not affected by the conscious realisation that there is no actual triangle.

When surgeons view images of the gallbladder and surrounding tissues on a two-dimensional screen, their unconscious brains automatically seek a pattern to match the mental model of the biliary tree stored in their long-term memory. But these are 'fuzzy' indistinct images. Object borders are obscured by blood and connective tissue. Being unconscious processes, these identifications are not open to introspective analysis. Unless powerful contradictory signs are found – and these are likely to be rejected or ignored by confirmation bias – the surgeons believe they have identified the cystic duct.

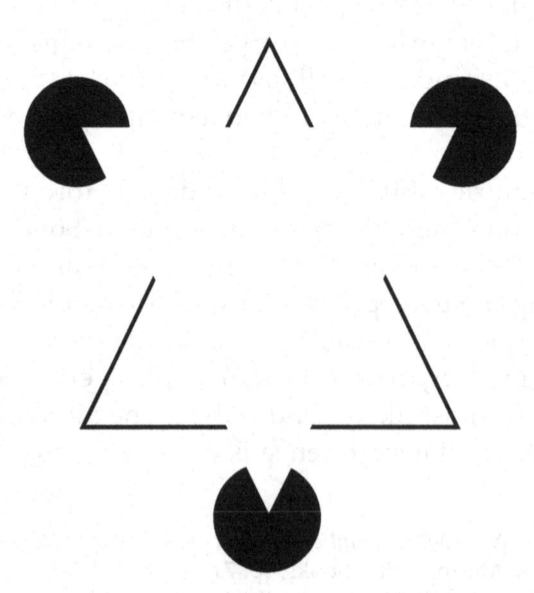

Figure 6.2 Kanizsa Triangle

Are such error traps inevitable? Way and his colleagues believe that additional training is unlikely to be the solution: these innate heuristics and biases cannot be eliminated, not least because they are highly adaptive and, for the most part, useful. They put their faith in technology, particularly visual techniques for verifying the anatomy (e.g., operative cholangiography).

Dr Gawande takes another stance: 'The statistics may say that someday I will sever someone's main bile duct, but each time I go into a gallbladder operation I believe that with enough will and effort I can beat the odds ... Operations like the lap chole have taught me how easily error can occur, but they've also showed me something else: effort does matter, diligence and attention to the minutest details can save you'.[22]

Wrong Site

Operations in which patients had the wrong body parts operated on have risen by a half in the period 2003–2006 within the UK. In 2003–2004, there were 27 claims for damages settled by the NHS Litigation Authority. In the following financial year, the number rose to 35 cases settled; and in 2005–2006, there were 40 cases settled. The cost of settling these claims increased from £447,694 in 2003–2004, to 663,145 in 2004–2005, and rose to over a million pounds in 2005–2006. A third of claims involved surgery on the wrong tooth, but wrong hips, knees and legs were operated upon.[23]

Dr Sally Giles and her colleagues[24] investigated experiences of wrong site surgery and current marking practices among 38 British consultant surgeons. Their specialties included ophthalmology, orthopaedics and urology. Most surgeons had experience of wrong site surgery, but there was no clear pattern of underlying causes. Marking the to-be-operated-on body part

---

22 Gawande (2002), p. 73. I have quoted this passage before but I do not apologise for doing so again because it has had such an enormous impact on my thinking and upon the structure of this book – though I do not necessarily agree that going the extra mile is the whole answer.

23 BBC News 24, 3 October 2006, 10:11 GMT.

24 Giles, S.J. *et al* (2006) 'Experience of wrong site surgery and surgical marking practices among clinicians in the UK.' *Quality and Safety in Health Care,* 15: 363–368.

varied considerably. Orthopaedic surgeons regularly marked before surgery; but some urologists and ophthalmologists said they did not. There seemed to be no formal hospital policies in place.

The United States has a similar perception of the rising incidence of wrong site operations – and this despite eight years of sustained patient safety efforts on the part of the Institute of Medicine and the Joint Commission on Accreditation of Health Care Organizations (JCAHO – pronounced 'Jayko').[25] In 2005, US healthcare facilities reported 84 operations to JCAHO that involved the wrong body part or the wrong patient, and this is likely to be an underestimate of the true incidence.

JCAHO requires hospital staff take an operating room 'time out' to verify the patient's identity, confirm the procedure to be performed and ensure that all the necessary equipment was present.

A recent study indicates that the problem might be 20 times more common than previously thought. It estimated that wrong-site surgery occurs between 1,300 and 2,700 times a year in the United States.[26] Reported cases included someone who had his sole functioning kidney removed and a stroke survivor who was supposed to undergo circumcision but had his testicles removed instead. Both patients had verbally verified the procedures beforehand.

The Elements of Recurrent Scenarios

In every recurrent accident scenario, there would seem to be at least three elements:

1. *Universals.* These are the ever-present hazards associated with a particular domain of activity. In the maritime world, for example, these would include rocks, shallows, currents and tides, and the presence of other vessels. In aviation they comprise gravity, weather and terrain. It is unplanned contacts with these universals that do the actual damage.

---

25 Davis, R. (2006) '"Wrong Site" surgeries on the rise.' *USA Today*, 17 April.

26 Seiden, S.C., and Barach, P. (2006) *Archives of Surgery*, 141 (September): 931–9.

2. *Local traps.* These are characteristics of the task or workplace that, in combination with human error and violation tendencies lure people into repeated patterns of unsafe acts or less-than-adequate performance. Such snares are likely to be fairly enduring features of a particular work situation. They are analogous to the snakes on the snakes-and-ladders board. If we land on a square with a snake's head then we are inexorably pulled down to the snake's tail. Translating this into the real world, each 'snake' in a hazardous workplace has a region of attraction around its head and the power to elicit a sequence of unwise acts along its body, while its tail leads into an area of unacceptable danger. The crucial feature of these 'snakes' is that they have the power to lure people into a series of unsafe acts, irrespective of who they are. Clearly, it is possible to resist these traps, but they nonetheless have a particular and persistent ability to lead people into danger.
3. *Drivers.* No psychologist was more concerned than Freud[27] with the motive forces that drive people into erroneous behaviour. In discussing the mechanisms underlying slips of the tongue, he made the following very pertinent observations:

> The influence of sound-values, resemblances between words, and common associations connecting certain words, must also be recognised as important. They facilitate the slip by pointing out a path for it to take. But if there is a path before me does it necessarily follow that I must go along it? I also require a motive determining my choice and, further, some force to propel me forward.

A very similar argument can be applied to the local traps in hazardous operations. Their mere existence is insufficient to explain why people are repeatedly – but not invariably – ensnared by them. They are the necessary but insufficient causes of recurrent accidents. The sufficiency is supplied by something that drives people towards and then along these treacherous pathways. The argument to be offered here is that, in hazardous work, this motive force is derived from an organisation's safety culture – or, more often, from the lack of it.

---

27 Freud, S. (1922) *Introductory Lectures on Psychoanalysis.* London: George Allen, (p. 36).

Cultural Drivers

It is clear from in-depth accident analyses that some of the most powerful pushes towards local traps come from an unsatisfactory resolution of the inevitable conflict that exists (at least in the short term) between the goals of safety and production. The cultural accommodation between the pursuits of these goals must achieve a delicate balance. On the one hand, we have to face the fact that no organisation is just in the business of being safe. Every company must obey both the ALARP principle (keep the risks as low as reasonably practicable) and the ASSIB principle (and still stay in business). On the other hand, it is now increasingly clear that few organisations can survive a catastrophic organisational accident.[28] But there are also a number of more subtle economic factors at work.

As Hudson has pointed out,[29] there can be a close relationship between the amount of risk taken and profitability. In hazardous work, as in exercise regimes, there is little gain without pain – or at least the increased likelihood of it. In oil exploration and production, for example, Hudson identified three levels of risk:

1. *Very low risk* where the return on investment may be only 8 per cent or lower – hardly more than would be expected from keeping the money in a bank.
2. *Moderate or manageable risk* where the return might be 12 per cent.
3. *High risk* where the return may increase to 15 per cent, but the margins between this and wholly unacceptable risks might be very small indeed.

To remain competitive, many companies must operate mainly in the moderate risk zone with occasional excursions into the high risk region. As the distance to the 'edge' diminishes so the number of local traps increases. Here, the 'snakes' are likely to be more numerous and need only be quite short to carry the system over

28 Reason, J. (1997) *Managing the Risks of Organizational Accident*. Aldershot: Ashgate.

29 Hudson, P.T.W. (1996) *Psychology and Safety*. Unpublished report. Leiden: Rijks Universiteit Leiden.

the edge, while the cultural drivers (pressures to get the job done no matter how) are likely to be exceedingly powerful indeed.

In summary, the same cultural drivers – time pressure, cost-cutting, indifference to hazards, the blinkered pursuit of commercial advantage and forgetting to be afraid – act to propel different people down the same error-provoking pathways to suffer the same kinds of accidents. Each organisation gets the repeated accidents it deserves. Unless these drivers are changed and the local traps removed, the same accidents will continue to happen.

Conclusion

One of the most important features of human fallibility is that similar situations provoke similar types of error and recurrent accident patterns involving different people. One of the main functions of an incident reporting system is to identify these recurrences and to indicate where remedial efforts must be directed. As we have seen in this chapter, recurrent accident patterns are not restricted to any one domain. Comparable patterns of repetition have been discussed in relation to railways, aviation, marine transport, nuclear power generation and health care. But these traps do not have to catch everyone. Cultural influences of one kind or another play an important role in springing the traps.

Chapter 7

Significant Accident Investigations

I have spent a large part of my working life studying accident and incident reports in a wide range of hazardous domains. They, perhaps more than any kind of material, have been the main basis for my thinking about the human and organisational contributions to the performance of complex, well-defended systems. In this chapter, I will examine three accident reports that have been especially significant in this regard.

I am also very interested in the *process* of accident investigation, particularly the snares, traps and pitfalls that lie in the path of those trying to reconstruct and understand past events. The accidents discussed here have been selected both for their theoretical importance and for the light they throw on investigative problems – though I will not be considering *hindsight bias* in any further detail: that was described earlier (Chapter 5) and has been extensively worked over by David Woods and his co-workers.[1] (An alternative title for this chapter could well have been 'Beyond hindsight bias.')

Problems with the Past

We can never recover the 'whole truth' about an accident, or indeed about any past event. The past is never completely knowable. Some of the reasons why this is the case are listed below:

- Although there may be 'hard facts' – of the kind captured by cockpit voice recorders and flight data recorders and from the pieces of wreckage that remain – the rest is inevitably best guess and theory.

---

1 Woods, D., Johannesen, L., Cook, R., and Sarter, N. (1994) *Behind Human Error: Cognitive Systems, Computers and Hindsight.* Wright-Patterson Air Force Base: OH: Crew Systems Ergonomics Information Analysis Center.

- Interpretations, particularly in regard to human actions, are subjective and shaped by the goals of the analysts and investigators.
- There are always alternative views. Some of these are rejected; others may be suppressed because they are socially or politically unacceptable.
- In reporting, we are 'digitising' complex interactive analogue events. I will expand on this point below.

All accident report writers have to chop up continuous and interacting sequences of prior events into discrete words, paragraphs, conclusions and recommendations. If we think of each sequence as a piece of string, then it is the investigator's job to tie knots in order to mark those points that appear to be critical stages in the development of the accident. Such identifications are necessary for making sense of the causal complexity, but they also distort the nature of the reality. However, if this parsing of events correctly identifies the proper areas for remediation and future accident prevention, then the problem is a small one; but it is important for those that rely on accident reports to recognise that they are – even the best of them – a highly selective version of the actuality. It is also a very subjective process.

As a university teacher, I have given successive generations of final year students the task of translating accident and incident narratives into event trees. Starting with the end event – the accident itself – they were required to track back in time, asking themselves at each stage what factors were necessary to bring about the subsequent events. Or, to put it another way, they were asked to consider which elements, if removed, would have thwarted the accident trajectory.

Even a simple starting narrative produced a wide variety of event trees, with different nodes and different factors represented at each node. While some versions were simply inaccurate, most were perfectly acceptable accounts. The moral was clear: the causal features of an accident are to the analyst what a Rorschach test (inkblot test) is to the psychologist's client – something that is open to many interpretations. The test of a good accident report is not so much its fidelity to the often-irrecoverable reality, but the degree to which it directs those who regulate, manage and operate hazardous technologies towards appropriate and workable countermeasures. In this respect, I believe that most professional accident investigators do a very good job indeed.

Changes in Accident Investigation

Over the past 50 years or so there has been a dramatic widening of the scope of accident investigations across many different hazardous domains. The nature of these changes is summarised in Figure 7.1.

In 1954, three de Havilland Comet aircraft – the world's first passenger jet airliner – crashed as the result of metal fatigue, all within a single year. It was natural, therefore, that air crash investigators focused closely on hardware problems. Metal fatigue was also implicated in the engine failure that stripped the United 232 of its three redundant and diverse hydraulic systems, and the fuselage break up on Aloha Airlines Flight 243, as well as contributing to the 1998 Eschede train disaster.

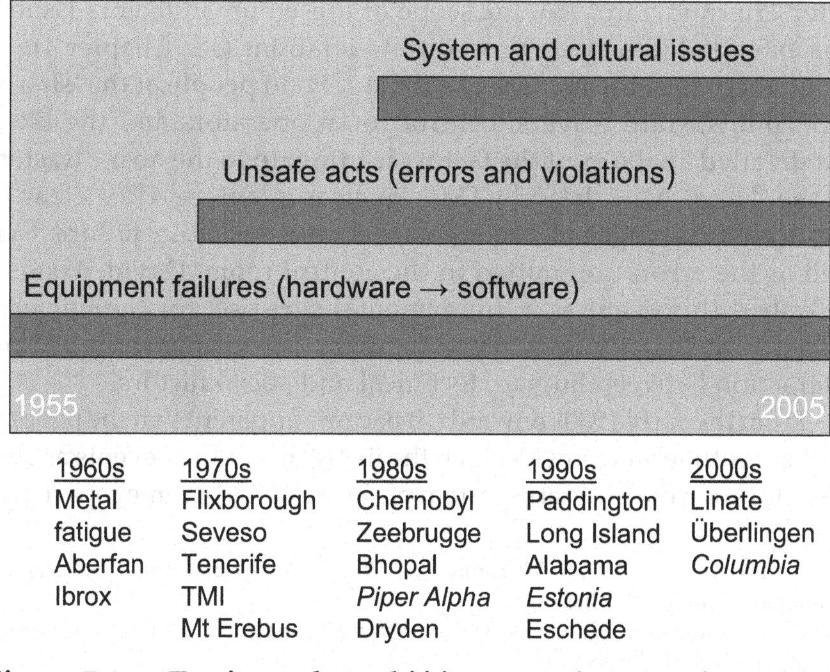

Figure 7.1 Tracing the shifting emphases of accident investigations

Clearly, concerns over hardware failures have not gone away, but some of the emphasis has shifted to software problems. In the early 1970s, Airbus Industrie broke into the Boeing-dominated aircraft market – very successfully, as it turned out. One of its innovations was the radical automation of the flight deck. The fact that human factors problems created by a flight management system could change modes both as the result of pilot action and according to its own internal logic played a significant role in the Airbus accidents of the 1980s.[2]

In the 1960s when the problem of error first began to attract attention, the estimated contribution of these human factors problems to transport accidents was around 20 per cent. In 1990, however, this estimate had increased fourfold to 80 per cent.[3] It wasn't so much that people had become more fallible as that greatly improved materials and engineering techniques had brought the human factor into greater prominence. In addition, the 1970s saw a number of very high-profile aircraft accidents, the most disastrous of which was the runway collision between two jumbo jets at Tenerife in 1977: the world's worst aircraft accident. After Chernobyl in 1986, the scope of these human factors issues was extended to include procedural violations (see Chapter 4).

In the 1970s, the focus was still largely on people at the 'sharp end': pilots, train drivers, control room operators and the like. But detailed analyses of the factors leading up to the near disaster at the Three Mile Island (TMI) nuclear plant in 1979 clearly implicated managerial, maintenance and systemic failures as well as the errors committed in the control room. David Woods[4] described this event as a 'fundamental surprise' for the nuclear industry in that its occurrence required an appreciation of the interaction between human, technical and social factors.

From the early 1980s onwards, it became apparent that the people on the frontline were not so much the instigators of bad events as the inheritors of workplace and system problems. This became especially

---

2 Reason, J. (1997) *Managing the Risks of Organizational Accidents*. Aldershot: Ashgate Publishing.

3 Hollnagel, E. (1993) *Human Reliability Analysis: Context and Control*. London: Academic Press (p. 3).

4 Woods, D. *et al.* (1994) *Behind Human Error: Cognitive Systems, Computers and Hindsight*. Wright-Patterson Air Force Base: OH: Crew Systems Ergonomics Information Analysis Center.

apparent with the capsize of the *Herald of Free Enterprise* in 1987[5] and the Mahon Report on the crash of an Air New Zealand DC10 into Mount Erebus in 1979[6] (to be discussed later in this chapter).

In April 1997, I attended a symposium organised by the US National Transportation Safety Board (NTSB) on 'Corporate Culture and Transportation Safety'. This event attracted some 550 delegates from all the NTSB's constituencies: aviation, railroads, highways, the marine world, pipelines and the safety of hazardous materials. The reason for the symposium was that NTSB accident investigators were increasingly conscious of the crucial role played by cultural factors in creating bad events.[7] Culture, unlike almost anything else within a system, can affect the integrity of the defences, barriers and safeguards for good or ill. We will revisit this notion in the concluding part of this book.

For the remainder of this chapter, I will deal with the seminal accident investigations thematically rather than chronologically, beginning with two legal inquiries that had an enormous influence not only on my own thinking but also on the way subsequent accidents are investigated.

The Mahon and Moshansky reports

Two accident investigations, in particular, had a profound influence on the development of the Swiss cheese model and the shift towards systemic explanations: the Mount Erebus DC10 crash in 1979 and the F-28 crash at Dryden, Ontario, in 1989. Each was the subject of a detailed public inquiry conducted by two eminent judges, the Honourable Peter Mahon in New Zealand and Mr Justice Moshansky in Canada.

One of the most significant features of the Mahon Report[8] was that it completely rejected the findings of an earlier investigation carried out by the chief investigator of New Zealand's Office of

---

5 Sheen, Mr Justice (1987) *MV Herald of Free Enterprise*. Report of Court No. 8074. Department of Transport. London: HMSO.

6 Mahon, Mr Justice (1981) *Report of the Royal Commission into the Crash on Mount Erebus, Antarctica, of a DC10 Aircraft Operated by Air New Zealand Limited*. Wellington: New Zealand.

7 Reason, J. (1998) 'Achieving a safe culture: Theory and practice.' *Work and Stress*, 3: 293–306.

8 Mahon, Mr Justice (1981).

Air Accident Investigations. This earlier report concluded that the crash was due almost entirely to flight crew errors. Not an unreasonable finding given that the sight-seeing DC10 crashed into a 13,000 foot Antarctic mountain at around 6,000 feet in Visual Flight Rules (VFR) conditions.

The Mahon Report, on the other hand, found no fault with the pilots. On the contrary, it concluded: 'The single dominant and effective cause of the disaster was the mistake by those airline officials who programmed the aircraft to fly directly at Mt Erebus and omitted to tell the aircrew.'[9] Mr Justice Mahon also expressed his view that Air New Zealand's case, as presented to the Royal Commission, had been an 'orchestrated litany of lies'; in short, a conspiracy to conceal the truth. Air New Zealand took this finding to the Court of Appeal who ruled that Mahon had no authority to make a conspiracy finding. Mahon appealed against this decision to the Privy Council who upheld the Court of Appeal judgement on the conspiracy issue, though it supported the Mahon Report's finding that the crew were not at fault.

The Mahon Report was ten years ahead of its time. Most of the accidents that have shaped our current thinking about organisational factors had yet to happen. The Moshansky Report,[10] however, was far more a product of its time in its sensitivity to the wide-ranging causes of organisational accidents.

The crash that it was set up to investigate had, on the face of it, all the hallmarks of a straightforward 'pilot error' accident. A commuter F-28 took off in snowy conditions without being de-iced and crashed a kilometre beyond the runway due the accumulation of ice on its wings. In another age, the flight crew's flawed decision to take off under such conditions could have been the end of the causal story.

But Mr Justice Moshansky interpreted his mandate very widely. In order to make the required recommendations in the interests of air safety, he felt it necessary 'to conduct a critical analysis of the aircraft crew, of Air Ontario [the carrier], of Transport Canada [the regulator] and of the environment in

---

9 Mahon, Mr Justice (1981) Vol 1, pp. 5–6.

10 Moshansky, Mr Justice (1992). *Commission of Inquiry into the Air Ontario Crash at Dryden, Ontario. Final Report.* Vol. 1, pp. 5–6. Ottawa: Ministry of Supply and Services.

which these elements interacted.' After considering the evidence of 166 witnesses, the 1700-page final report implicated the entire Canadian aviation system:

> Had the system operated effectively, each of the [various causal] factors might have been identified and corrected before it took on significance. It will be shown that this accident was a result of a failure in the air transportation system as a whole.[11]

The need for air accident investigators to take account of the wider managerial and systemic issues was formally endorsed by the International Civil Aviation Organization in February 1992. The minutes of the ICAO Accident Investigation Divisional Meeting reported the following discussion item:

> Traditionally, investigations of accidents and incidents had been limited to the persons directly involved in the event. Current accident prevention views supported the notion that additional prevention measures could be derived from investigations if management policies and organisational factors were also investigated.

The meeting concluded by recommending that a paragraph be inserted in the next edition of ICAO Annex 13 – the standards and recommended practices for air accident investigators throughout the world. It appeared in the Eighth Edition of Annex 13 (Aircraft Accident and Incident Investigation, 1994) as follows:

> 1.17. Management information [that accident reports should include]. Pertinent information concerning the organisations and their management involved in influencing the operation of the aircraft. The organisations include, for example, the operator, the air traffic services, airway, aerodrome and weather service agencies, and the regulatory authority. The information should include, but not be limited to, organisational structures and functions, resources, economic status, management policies and practices, and regulatory framework.

A short while after the publication of this ICAO Annex, an accident investigation adopting this particular recommendation played a significant part in producing a radical change in the Australian civil aviation system. The Bureau of Air Safety Investigation cited regulatory surveillance failures by the Australian Civil Aviation Authority (CAA) as having contributed to the crash of a small commuter aircraft at Young, New South Wales, in June 1993. As the result of this finding and other factors, the then

11 Moshansky, Mr Justice (1992), pp. 5–6.

(Liberal) government disbanded the Australian CAA in 1995, and later reconstituted it as the Civil Aviation Safety Agency.

Has the Pendulum Swung Too Far?

I have traced the ever-widening spread of causal fall out from the time when it was sufficient to blame the people immediately concerned, through various organisational and systems issues, to the sacking of a Civil Aviation Authority in Australia. But has the process gone too far towards collective responsibility and away from individual responsibility?

Models of accident causation can only be judged by the extent that their applications enhance the safety of transport. The economic and societal shortcomings, identified along the way, are beyond the reach of system managers. From their perspective, such problems are given and immutable – at least until the next election. Accident investigations have three related goals: explanation, prediction and countermeasures. As I see them, the relative contributions of individual factors, workplace factors, organisational processes and culture, regulation and societal issues are summarised below:

- Individual factors alone have only a small to moderate value for all three goals.
- Overall, workplace and organisational factors contributed the most added value.
- There are rapidly diminishing returns on pursuing the more remote influences, particularly in regard to countermeasures and risk management.

Remote factors have little causal specificity. Like a shotgun blast, the more remote the firing, the greater will be the spread of the influences. Their impact is felt by many systems. Such influences are outside the control of system managers and, from their point of view, are largely intractable. The more exhaustive the inquiry (e.g. the Moshansky Inquiry), the more likely it is that it will find remote factors and a large number of latent 'pathogens'. But their presence does not discriminate between normal states and accidents – either within the same organisation at different times, or between different organisations in the same transport domain. Only the local triggers and proximal factors can do that. As such they are more properly conditions rather than causes.

Conditions and Causes

Hart and Honoré,[12] writing on causation in the law, provided a very clear distinction between conditions and causes.

> These factors [conditions] ... are just those which are present alike both in the case where such accidents occur and in the normal cases where they do not; and it is this consideration that leads us to reject them as the cause of the accident, even though it is true that without them the accident would not have occurred . . . To cite factors that are present in the case of disaster and normal functioning would explain nothing... Such factors do not 'make the difference' between disaster and normal functioning.

Some examples of conditions and causes are given in Table 7.1.

Counterfactual Fallacy

One of the things I have noticed in recent accident reports and particularly that of the Columbia Accident Investigation Board (CAIB) is the appearance of the counterfactual fallacy. Here are two examples from the CAIB report:

> In our view, the NASA organizational culture had as much to do with this accident as the foam. [You will recall that a large lump of foam penetrated the leading edge of the Shuttle's wing]

Table 7.1 Three examples of the distinction between cause and condition

Conditions	Causes
FIRE: Oxygen, dryness, etc.	Sources of ignition.
IRISH POTATO FAMINE: Amount of rainfall. Potato the sole crop.	Potato blight fungus.
TRAIN CRASH: Normal speed, load and weight of the train. Routine stopping and acceleration, etc.	Collision, explosion, derailment and the like.

12 Hart, H.L.A., and Honoré, T. (1985) *Causation in the Law.* Second edition. Oxford: Clarendon Press, (p. 34).

> The causal roots of the accident can be traced, in part, to the turbulent post-Cold War policy environment in which NASA functioned during most of the years between the destruction of *Challenger* and the loss of *Columbia*.[13]

During these years, NASA was directed by a cost-cutting, down-sizing, outsourcing, 'leaning and meaning' administrator. But these were the buzz words of the age. Everyone was exposed to them. They are conditions not causes.

The counterfactual fallacy goes as follows. All accident investigations reveal systemic shortcomings. They are present in all organisations. It is then but a short step to argue that these latent conditions caused the accident. There are always organisational interventions that could have thwarted the accident sequence, but their absence does not demonstrate a causal connection. So the fallacy is this: if things had been different then the accident would not have happened; *ergo* the absence of such differences caused the accident. But, as stated above, these organisational factors are conditions not causes.

The Current View

Accidents happen because ...

- *Universals*: the ever-present tensions between production and protection[14] create
- *Conditions*: latent factors that collectively produce defensive weaknesses that
- *Causes*: permit the chance conjunctions of local triggers and active failures to breach all the barriers and safeguards.

The 'universals' and 'conditions' are ubiquitous. Only the local events are truly causal and 'make the difference' between this accident and all the other organisations (or the same at a different time) that remain accident-free.

So what are the 'conditions' in organisational accidents? They are the 'usual suspects' – the latent pathogens present in virtually all hazardous systems: inadequate tools and equipment, a less-

13 Columbia Accident Investigation Board (2003) CAIB Report Volume 1. Washington DC: Government Printing Office, (p. 178).

14 These tensions were discussed at some length in Reason, J. (1997).

than-adequate safety culture, poor design and construction, workarounds, management and supervisory shortcomings – and so on, and so on.

Have we come full circle? Am I advocating a return to the blame culture of 1950s and 1960s? No, the fact that latent conditions are everywhere does not diminish management's responsibility to identify and rectify them. Making regular checks on the 'vital signs' of an organisation is what management is paid to do: it is not safety specific, it's just good business.

PART IV
Heroic Recoveries

Chapter 8

Training, Discipline and Leadership

This chapter presents two military case studies, both of them involving retreats undertaken under extremely hazardous circumstances.

The first one occurred during the Peninsula War at the Battle of Fuentes de Onoro in Spain near the Portuguese border. This was a three-day battle between a British-Portuguese army and the French Army of Portugal in early May 1811. Wellington's forces (then Sir Arthur Wellesley, Viscount of Wellington) eventually won – though, as he later said of Waterloo, it was 'a close run thing' not least because he made a serious error in leaving his right flank exposed.

The second retreat happened during the Korean War in the winter of 1950 when a very large number of Chinese soldiers crossed the Manchurian border into North Korea at the Chosin Reservoir and entered the war against the United Nations forces. In this locality, the UN forces were largely made up of an American infantry division and the 1st Division of the US Marines. Both units were forced to retreat southwards – though the marines preferred to describe it 'as attacking to the rear'. The subsequent performance of these two units differed dramatically, as we shall see later.

The Light Division's Retreat at Fuentes de Onoro (1811)

Before describing this event, it is worth mentioning something about the nature of battles in the eighteenth and early nineteenth centuries. Armies had three kinds of combat soldiers: infantry, cavalry and artillery. The basic rules of war in those days were fairly simple; as in the children's game of stone, scissors and paper, outcomes depended on who was matched against whom.

Cavalry beat infantry when the latter were disorganised or in column; infantry beat cavalry when the former were assembled into squares. Artillery, of course, destroyed both with equal ease.

Forming Square on the March

Figure 8.1 shows the basic elements of switching from marching to forming square. Let us imagine that this is a battalion with four companies (referred to in the drill book as divisions). Each company is between 12 and 20 men wide and four rows deep.

On the colonel's command 'Form square on the first division', the major commanding the leading company (#1) orders it to halt. The sergeants shout 'dress ranks' and the men shuffle closer together looking to their right to ensure that their line is straight. While this is going on, the second and third companies receive the following order 'Outward wheel. Rear sections close to the front' – these instructions are then echoed by the section sergeants. The left-hand sections of the 2nd and 3rd companies swing through 90 degrees to the left while the right-hand sections swing to

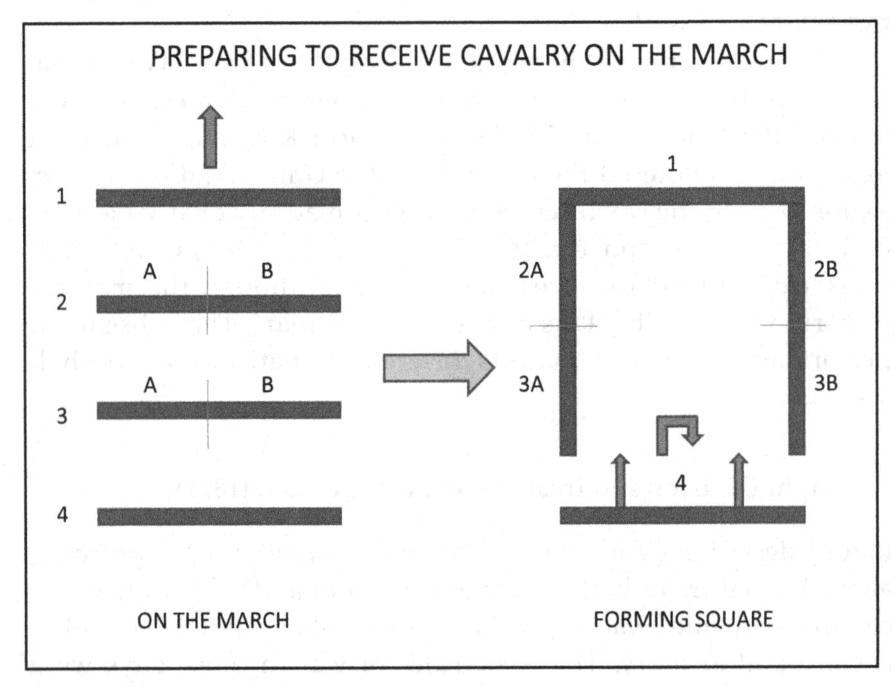

Figure 8.1 The manoeuvres required in preparing to receive cavalry on the march

the right to form the two sides of the square, each side facing outwards. As they swing around, the innermost men reduce their marching pace from 30 to 20 inches, while those on the outside with furthest to go increase their stride to 33 inches. When these manoeuvres are completed, the fourth company marches briskly forward to form the rear of the square, and then turns about so that it too is facing outwards. Meanwhile the officers ride their horses into the square and take up their respective stations along with the colour party (two ensigns and a number of sharpshooters to protect them) and any supernumeraries.

Once the squares are formed, the front ranks kneel and dig the butts of their muskets into the ground with the bayonets facing obliquely outwards. When the command 'prepare to receive cavalry' is given, the second and third rows aim their loaded muskets at the advancing horsemen, while the fourth rank holds their weapons at the port ready to fill gaps as they arise.

It will be appreciated that these realignments require impeccable timing on the part of the colonel and the company commanders. If enemy cavalry got inside the squares while they were forming, a massacre would ensue.

Although highly vulnerable to artillery, infantry squares were proof against most cavalry attacks. Squares owed their relative impregnability to a very understandable quirk of equine psychology: horses cannot be persuaded to charge walls of bristling bayonets. They veer away at the last minute, often propelling their riders into the square. When the cavalry have withdrawn, the order to march is given and the columns re-form in reverse order and move away. Although pursued by a large body of French infantry, the Light Division outdistanced them since they marched at 108 steps per minute, faster than any other infantry units of the time.

In the event of this particular battle, the Light Division's manoeuvres were a little more complicated than those described above, though the principle was the same. Each battalion had ten rather than four companies. When ordered to form square, the first company would halt and the second one would march smartly to its right, forming a continuous line of up to 40 men in four ranks as the front side. The six centre companies swung left and right to form the sides of what was in fact an elongated rectangle rather than a square. The last two companies, side by side, filled in the empty space at the back. Though, since they were retreating northwards towards Wellington's main army around the village of Fuentes de Onoro, the south or back end of the square was the side directly facing the enemy.

The Campaign and the Battle

The French Army of Portugal, commanded by Marshal Masséna, had spent a miserable winter outside the Lines of Torres Vedras that Wellington had constructed to defend Lisbon. Masséna failed to breach this double line of interlocking fortifications. He withdrew his starving army to the Spanish-Portuguese border with Wellington's Anglo-Portuguese army following close behind.

With Portugal now secured, Wellington set about retaking the fortified frontier cities of Almeida, Badajoz and Ciudad Rodrigo. While Wellington was besieging Almeida, Masséna marched to relieve the French garrison in the city. Wellington chose to thwart this relief attempt at the small village of Fuentes de Onoro with an army of 38,000 men (comprising 36,000 infantry, 1,850 cavalry and 48 guns).

On 3 May, Masséna launched a frontal assault against the British regiments holding the barricaded village. He had an army of 47,000 troops (42,000 infantry, 4,500 cavalry and 38 guns). The village was the centre of the fighting for the whole of the first day of this three-day battle, the longest engagement of the Peninsula War. Initially, the French drove the British back under tremendous pressure, but by evening the defenders had reclaimed the streets and buildings lost earlier in the day. The French lost 650 casualties against only 250 British losses.

The next day saw little fighting. Both sides rested from the ferocity of the previous day's combat. However, a French reconnaissance patrol discovered that Wellington's right flank was over-extended and rested on the weakly held hamlet of Poco Velho, about a mile to the south of Fuentes de Onoro. Wellington had left the 7th Division exposed and in danger of being cut off from the bulk of his army in and around Fuentes de Onoro. A sketch of the battlefield on 5 May is shown in Figure 8.2.

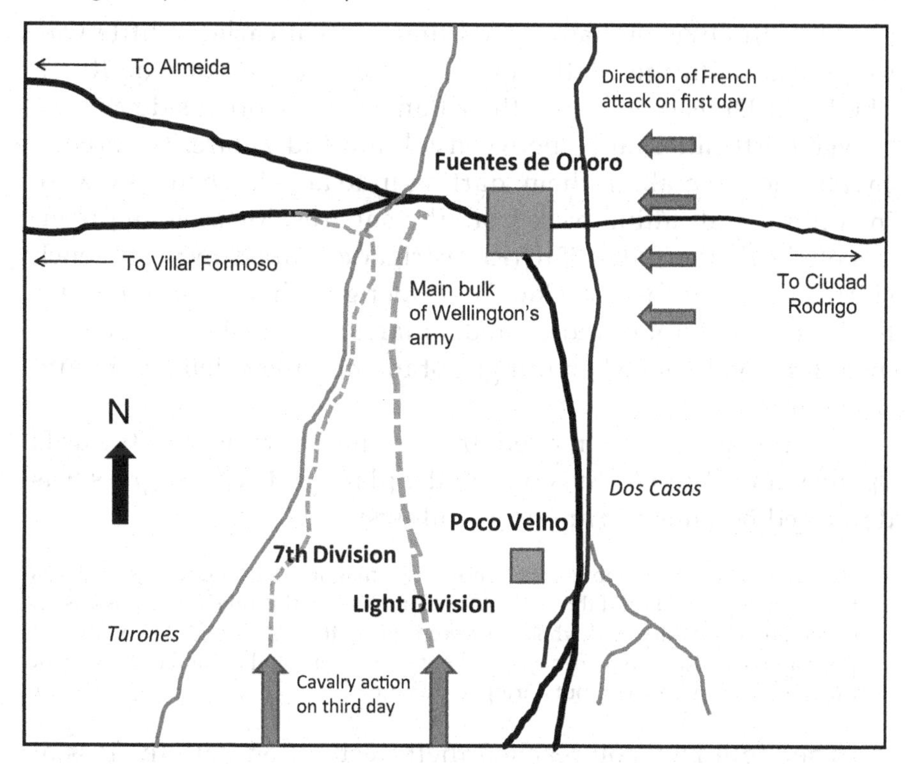

Figure 8.2 The Battle of Fuentes de Onoro, 5 May 1811

At dawn on 5 May, Masséna launched a heavy attack on the weak British right wing. It was led by dragoons with some infantry divisions in support. Almost immediately, two of the 7th Division battalions were badly mauled by French cavalry. Wellington acted quickly to save the 7th Division from annihilation, sending the Light Division, under General Craufurd (who had returned from home leave the previous day), to cover the withdrawal of the 7th Division by drawing off the enemy's troops. They were supported in this by British cavalry and the King's German Legion, around 1,500 horsemen as opposed to the 3,500 cavalry fielded by the French.

Between the Light Division and the main body of the Anglo-Portuguese Army lay a scrubby plain swarming with French cavalry. As they emerged from under the trees, they formed up in columns of companies and, as they came into the open ground, they closed the intervening gaps so that there were only about 15 feet between the fourth rank of one company and the first rank of the next.

The 7th Division had started their retreat earlier a little way to the east with their left flank protected by the Turonne River. The Light Division covered their flanks on the open side as they moved north, alternating between column and square. The French cavalry hovered about them, darting in to attack when they were in column and falling back from the squares. Meanwhile green-jacketed riflemen of the 95th (Rifle Brigade) sniped away at French skirmishers and the occasional galloper gun that was brought up to fire on the largely red-coated squares. Skirmishers were easy meat for cavalry so when things got too hot, the riflemen ran into the squares.

In this way, they moved three miles northwards towards Spencer and Picton's divisions on the plateau. Their progress was described by an eye-witness[1] as follows:

> 'The execution of our movement presented a magnificent military spectacle, as the plain between us and the right of the army was by this time in the possession of the French cavalry, and while we were retiring through it with the order and precision of a common field-day, they kept dancing around us, and every instant threatening a charge, without daring to execute it.'

The Light Division reached their destination with the loss of only 47 men. The French cavalry suffered many more casualties. The British and German cavalry had repeatedly charged the much larger numbers of French horsemen during the retreat. It was truly a testimony to the effective combination of the two arms: infantry and cavalry.

Bernard Cornwell, as ever, describes their arrival very eloquently:

> 'The French horsemen could only watch their enemy march away and wonder why in over three miles of pursuit across country made by God for cavalrymen they had not managed to break a single battalion.' [2]

Neither General Craufurd nor Lord Wellington made much of this extraordinary feat by the Light Division. Forming square on

---

1 Captain Sir John Kincaid (1830) *Tales from the Rifle Brigade: Adventures in the Rifle Brigade and Random Shots from a Rifleman*. Barnsley: Pen and Sword Military (reprinted 2005), (p. 38). Captain Kincaid was by no means the only rifleman to record his experiences in the Peninsula War. Others included Rifleman Costello and Major George Simmons. But perhaps the best description of the retreat is provided by Bernard Cornwell in his novel *Sharpe's Battle*.

2 Cornwell, B. (1995) *Sharpe's Battle*. London: Harper-Collins, (p. 352).

the march is just one of the many manoeuvres that these soldiers had practised for hours and hours on the parade grounds of their home depots. They were professionals. Cavalry had been a well-understood hazard for over two hundred years. Defending against it was what they had been trained to do.

In his dispatch to Lord Beresford on 8 May 1811, Wellington covered these events in a single short paragraph:

> The movement of the troops upon this occasion was well conducted, although under very critical circumstances, by Major General Houston, Brig. General Craufurd, and Lieut. General Sir Stapleton Cotton. The 7th division was covered in its passage of the Turon by the Light Division under Brig. General Craufurd; and this last, in its march to join the 1st division, by the British cavalry.

In the end, Wellington was victorious, but he never counted it among his victories. Masséna's main aim was to secure the village of Fuentes de Onoro. But he did not succeed. After spending three days bombarding and parading before the British position, Masséna gave up the attempt and withdrew. The French losses varied, according to the sources, from 2,200 to 3,500. The Anglo-Portuguese losses were in the region of 1,500. Wellington acknowledged how precarious his position had been by saying later: 'If Boney had been there, we would have been beat.' Wellington was able to resume his siege of Almeida, and the battle was won.

The Withdrawal of the 1st Marine Division from Chosin Reservoir (1950)

Background

Korea, a peninsula the size of Florida, was divided (and still is) into two countries at the 38th Parallel. The American 8th Army (initially four under-strength divisions previously stationed in Japan) had travelled a long way since late June 1950 when they came to the aid of the Republic of Korea Army (ROK) that was under assault from ten divisions of the formidable North Korean People's Army (NKPA), equipped with the latest Russian military technology. For the next two months, they retreated southwards. By August, the American forces (under the nominal command of the United Nations) and their ROK allies occupied a stretch of land some 75 by 40 miles in area around the South Korean port of Pusan at the tip of the peninsula. They had suffered extensive losses in the summer fighting.

In September 1950, they were massively reinforced both in land forces and air power. The first non-American troops had arrived and the entire force was led by General MacArthur, Commander in Chief Far East Command. The 8th Army and its allies held the North Koreans in check while a new force, X Corps, was formed. This was spearheaded by the 1st Marine Division. It also included the 7th Infantry Division and some ROK units. They made a successful landing at Inchon high up on the western coast of South Korea and linked up with the 8th Army near Osan South Korea.

By the end of September, the UN forces had liberated Seoul. In mid-October they captured Pyongyang, the North Korean capital and resolved to move north to the Yalu River on the Manchurian border. X Corps had withdrawn from Inchon harbour and was subsequently landed on the east coast of North Korea near Wonsan Harbour. In late October and early November, X Corps and the 8th Army had a taste of things to come when the Chinese Revolutionary Army attacked from the hills in what they called their First Phase Offensive.

On 27 November, a Chinese army of some 60,000 men poured over the Manchurian border. Their aim was to destroy a force of 12,000 US Marines, marching north to the Yalu River. Three marine regiments were strung out along 80 miles of a narrow mountain road that snaked its way up to the high plateau of the Chosin Reservoir. A merciless wind drove temperatures down to 30 below. Soon the marines were surrounded by eight Chinese divisions who emerged at night to pounce on the unsuspecting Americans. The Marines suffered serious losses, but for five days and nights they fought off waves of bugle-blowing and drum-beating Chinese infantry. They inflicted huge casualties upon the Chinese, both then and during the subsequent withdrawal to the port of Hungnam. Figure 8.3 shows a sketch of the area over which they retreated and fought.

The Forces Engaged in the Chosin Campaign

X Corps X Corps, led by Major General Ned Almond,[3] was primarily made up of American troops – though there were 300

3 A controversial officer much distrusted by the Marines.

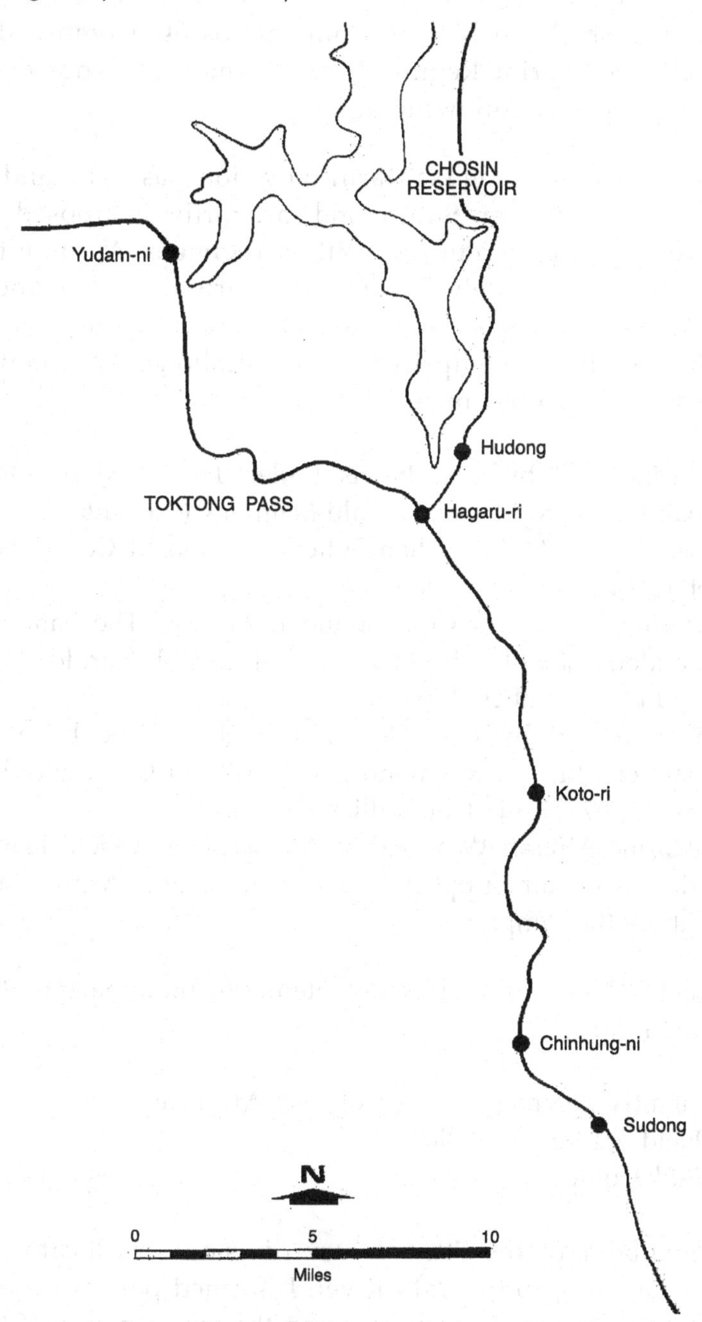

Figure 8.3 Showing the area over which the Chosin Reservoir campaign was fought

men of the British Royal Marine Commandos (41 Commando) attached to the 1st Marine Regiment. For the most part, however, X Corps comprised the following units.

1st US Marine Division The 1st Marine Division was commanded by Major General Oliver Smith and primarily composed of three infantry regiments plus an artillery regiment. An infantry regiment has approximately 3,500 men – virtually all of them combat soldiers (the Navy supplied the medical corpsmen). Each infantry regiment comprised three battalions. An infantry battalion has around 1,000 men.

- 1st Marines led by Col. Lewis Puller. His three battalion commanders were Lt Col. Donald Schmuck (1st Battalion, 1st Marines or 1/1); Lt Col. Allan Sutter (2/1) and Lt Col. Robert Taplett (3/1).
- 5th Marines led by Lt Col. Raymond Murray. The battalion commanders: Lt Col. John Stevens (1/5), Lt Col. Harold Roise (2/5) and Lt Col. Robert Taplett.
- 7th Marines led by Col. Homer Litzenberg. The battalion commanders: Lt Col. Raymond Davis (1/7), Lt Col. Randolph Lockwood (2/7) and Lt Col. William Harris.
- First Marine Aircraft Wing led by Major General Field Harris provided close air support to the Marine and Army units throughout the campaign.

7th Infantry Division The following elements took part in the Chosin campaign:

- 31st Infantry Regiment led by Col Allen MacLean.
- 57th Field Artillery Battalion.
- 31st Tank Company.

The remainder of the 7th Infantry division was located to the north-west, close to the Yalu River. It formed part of the 8th Army, that had advanced northwards on the western side of the Korean peninsula

The Chinese People's Liberation Army (PLA)

The second Chinese offensive began on 26–27 November. It comprised some 300,000 PLA soldiers. Eighteen lightly armed divisions attacked the 8th Army on the Yalu River in the west and another 12 divisions were thrown against X Corps around the Chosin Reservoir. Opposing them were 17 8th Army and X Corps divisions. In addition, there were high-quality British, Commonwealth and Turkish troops whose collective strength amounted to another division. If one counts in the remaining North Korean troops who lent some support to the Chinese, the opposing forces were roughly equal in size – some 400,000 men each. The Chinese had enormous reserves behind them in Manchuria, but the UN forces, at least in the north-eastern sector, had a huge superiority in material: artillery, transport, food, cold-weather clothing, radios, medical facilities and air superiority that was only occasionally challenged by MIG-15s based in Manchuria and flown by Chinese (and possibly Russian) pilots.

Far East Command and UN intelligence services underestimated the strength of the PLA forces opposing them – which was understandable given the difficulties of identifying PLA units as they crossed the border. They came at night and were largely on foot. A graver error, however, was the assumption that Chinese soldiers were an inferior version of the North Korean troops that they had been fighting (and latterly beating) over the past six months. Whereas the North Koreans had modern Soviet weapons and tanks, the Chinese possessed only a motley collection of captured Japanese and American weapons. A Chinese division had barely nine pieces of artillery, and these were mainly light 76-mm howitzers and no trucks. But they had formidable strengths to compensate for these limitations.

The Chinese did not fight like the North Koreans, who depended upon frontal assaults backed by T-34 tanks. The PLA attacked mainly at night, using large quantities of hand grenades, sub-machine guns (including Tommy guns) and mortar fire at very close range – usually from the rear of the UN positions. Although their cotton quilted jackets gave little protection from the cold, they were double-sided, green on the outside, white on the inside. When the snow came, they reversed their jackets and so remained well camouflaged. They were masters of concealed movement and surprise, often coming very close to Marine positions before they rushed forward in waves, blowing bugles, banging drums, throwing grenades and spraying bullets from their light machine guns; though, this was mostly a suicidal tactic when faced by determined Marines, well dug in and using .50 calibre heavy machine guns.

The Fighting

The Chosin campaign was fought in a ten-day period between 27 November and the 6 December, 1950. Although the fighting was continuous, it can conveniently be divided in four different actions. Figure 8.3 shows the locations of these battles, and the route of the Marine withdrawal:

- The successful defence by the Marines of Hagaru-ri.
- The successful defence by the Marines of Yudam-ni.
- The successful breakout of the 5th and 7th Marines from Yudam-ni to Hagaru-ri.
- And the effective destruction of a US Army task force (RCT 31) from the 7th Infantry Division on the eastern side of Chosin Reservoir.

The 8th Army, retreating on the west side of North Korea suffered the worst disaster to befall American forces since the Battle of the Bulge in 1944. In Korea, as in the Ardennes in World War Two, whole divisions of the US Army were rendered 'combat ineffective'. But for the Chinese it was a Pyrrhic victory, thanks very largely to the fighting withdrawal of the 1st Marine Division.

For them, the retreat could be described as a victory. The fighting fit brought back their wounded and a large number of frozen dead. Over 1000 marines were killed or reported missing, and 3,500 were wounded. Almost twice as many suffered frostbite. But they effectively destroyed six of the eight Chinese divisions that surrounded them. China's 3rd Field Army did not reappear in the war until March 1951. Their exact losses are uncertain, but captured Chinese documents indicated that some 25,000 men died in the fighting, another 12,000 were wounded and tens of thousands more fell victim to the freezing weather. When X Corps evacuated eastern Korea from the port of Hungnam in December, they were under no significant Chinese attacks. The bulk of the Chinese light infantry were lying on the sides of the road or around American defensive position on the surrounding hills. So how did they do it?

What Gave the US Marines the Edge?

There were many factors that contributed to the successful withdrawal of the X Corps, but I will focus upon the 1st Marine Division's contribution. The main points of difference are listed below:

- Like their Army counterparts on the western side of the peninsula, most of the marines doing the fighting in north-east Korea were young. Here is a description of them from one of their number:[4] 'Many of these kids had just started shaving. They were ordinary-looking teenagers, maybe a bit more hard-bitten than most. Journalists referred to them as men, but if you accompanied them on a Stateside liberty you'd notice they bought more candy and ice cream than beer or whiskey, and they were pretty bashful around girls. They would call the girls' mothers "Ma'am". I'm talking about the privates and the privates first class (PFCs) who made up most of the Corps and did most of the fighting.'
- Unlike their Army counterparts, Marine recruits were inspired from the beginning with the unshakeable belief that they had joined a select and elite body of men. The job of Marines was to capture beachheads and hold them until the occupation troops arrived – usually GIs (Army soldiers). As Martin Russ[5] put it: 'The ultimate payoff of this esprit de corps was a headlong aggressiveness that won battles.'

---

4 The description is by PFC Ray Walker of the 5th Marine Regiment. It is quoted in Martin Russ's excellent book: *Breakout: The Chosin Reservoir Campaign, Korea 1950.* (New York: Fromm International Publishing Corporation, 1999, p. 7). This book is a detailed record of the campaign from someone who was there.

5 Ibid, p. 5.

- Recruits to the US Marine Corps were initially trained at either Parris Island (in the east) or San Diego (in the west). They endured ten weeks in a hostile and wholly enclosed environment. There, the recruit was 'subject to such harassment and confusion – very much like combat itself – that it spawned in his homesick heart a desperate yearning for order, and finally a love of that order and a clear understanding that in its symmetry lay his safety and survival.'[6]
- Each Marine rifle regiment was made up of well over 3,000 actual fighting men. As mentioned earlier, their medical corpsmen came from the Navy. Unlike the Army, there were very few rear-echelon characters like the streetwise Sergeant Bilko and his cronies whose nefarious and largely non-military antics amused TV viewers in the 1950s and 1960s. And, unlike the 8th Army, each Marine rifle regiment was at full strength at the beginning of the campaign. So under-strength was the US Army at the start of the war that the high command instituted the KATUSA programme (Korean Augmentation of United States Army). This meant that the 8th Army's diminished numbers were fleshed out with South Korean troops. This was a failure on many counts. In their excellent book, *Military Misfortunes,* Cohen and Gooch described the KATUSA situation: 'Not surprisingly, barriers of language and culture prevented the ill-trained KATUSAs from becoming fully fledged members of their host units. This lack of mutual trust proved disastrous, particularly in night combat, when American troops and KATUSAs had no confidence, or, indeed, intelligible conversation with one another. In November 1950 the KATUSA program reached its Korean War peak: 39,000 Korean troops served in American units.'[7] The only division that stood at full strength and had no KATUSAs was the 1st Marine Division. My point here is that many of the factors that brought about the stark differences between marines and GIs were systemic, organisational and political in origin. Many 8th Army units fought gallantly and effectively, particularly the remnants of RCT 31 who protected the right of the X Corps withdrawal.

6 Ibid, p. 5.

7 Cohen, E.A., and Gooch, J. (1991) *Military Misfortunes: The Anatomy of Failure in War*. New York: Vintage Books, (p. 183). This is a primary source for those who wish to gain a broad-spectrum view of the campaign.

- Other systemic advantages included the following:
  - The Marines got closer and more persistent air support from their own air wing than did the Army units.
  - General Smith, the divisional commander, insisted on slowing his advance northwards in order to establish adequate airstrips and supply dumps along the main supply route. These logistical measures ensured that the 1st Division was well supplied with the necessities of war – ammunition, plasma, food, warming tents, winter weather clothing, barbed wire and trip flares – throughout the campaign. They also had relatively easy access to Navy supply ships operating to and from nearby ports. The 8th Army was not so fortunate. Almost every soldier got a Thanksgiving dinner with turkey and trimmings, but this required the halting of airlifts of winter field jackets from Japan. They were also short of gasoline, grenades, trip flares, barbed wire and antipersonnel mines – though they rarely asked for these last four items to be replenished, even though they would have proved very useful in the kind of combat they were facing.
- Marines also benefited from taking basic precautions that were neglected by the Army units:
  - They dug in every night and set up barbed wire, trip flares and noisemakers.
  - They employed aggressive patrolling so that they could seize the high ground wherever possible.
  - Battalion commanders checked that troops had dug foxholes before sleeping, even though this meant chipping away at a foot of frozen dirt.
  - Company commanders made sure that each man changed his sweat-soaked socks every evening – despite the freezing cold – in order to prevent frostbite.
  - Exhausted at the end of their retreat, they were nevertheless ordered to clean their weapons and all other gear before boarding the evacuating ships. This was in keeping with their basic training.
- The Marine officers had learnt some important tactical lessons during the first Chinese offensive. To deal with the Chinese practice of attacking at night, they ordered that the defending

units had only to maintain position until daylight. When vision was restored, Marine firepower would destroy the massed Chinese troops. This is what they did with great success, despite large-scale penetrations and infiltrations of their defensive positions.

- Above all, the Marines were fired by an unrelenting offensive spirit. Unlike the isolated companies of the 2nd Infantry Division, they did not immediately fall back under the pressure of the Chinese attacks. Where possible, they dug in on the high ground within larger defensive perimeters than those used by the 8th Army.
- Since the regiments of the 1st Marine Division were distributed over a fairly large area around the Chosin Reservoir, the fighting was coordinated by the regimental and battalion commanders. Company and platoon leaders also played a vital role. Some were so aggressive that they even frightened the marines under their command. One such individual was 1st Lt Chew-Een Lee, a Chinese-American officer, and one of the very few people in the UN forces who understood what the enemy was saying. His greatest fear was that he would be removed from his fighting platoon and transferred to an intelligence unit. This did not happen and Lt Chew-Een Lee fought furiously throughout the campaign, even though wounded more than once.

So what were the main ingredients of the 1st Marine Division's heroic recovery from the unexpected onslaught of huge numbers of Chinese troops? They were not supermen: for the most part, they were teenagers hardly any different from their GI counterparts. Nor were they all regulars. Approximately half the division were reservists, called up a short time earlier from their everyday civilian lives – although most of these had fought the Japanese in the Pacific during World War Two. Nor did they have more Korean War combat experience than the 8th Army, some of whose number had been fighting in Korea for six months – though there were not so many of these combat veterans left.

So what gave them the edge? *Esprit de corps* was clearly a very important contributor. This, in practical terms, is the unswerving conviction that it is better to die than to leave your comrades in the lurch; that every Marine had to live up to the distinguished battle record of the Corps; that teamwork, discipline, courage and the spirit of the offensive were the prerequisites of victory. But none of these things was necessarily unique to the Marine Corps, though they may have had these qualities to a greater and more sustained degree than their Army colleagues – thanks to the intensive training they had received at the Parris Island and San Diego basic training camps.

I believe that their success in Korea was derived in large part from three rather more basic factors. First, a strict but not mindless adherence to 'by-the-book' methods that covered all aspects of life on the battlefield, from the drudgery of changing socks to fighting tactics. And the 'book' in this instance was not written in stone, it was highly adaptive and tailored specifically to the local conditions. Second, the 1st Marine Division was, in this campaign (though it may not always have been the case) extremely fortunate in the quality and fighting skills of it officers at all levels. Third, there was the excellent support that the frontline marine received from the Corps system at large: the Navy, the logistics, its air wing and artillery. As we shall see again and again throughout this section, systemic issues are as important to heroic recoveries as they are to the creation of unsafe acts.

Concluding Remarks

Both of the divisions discussed in this chapter – Wellington's Light Division and the US 1st Marine Division – were elite formations. And this would have counted for good deal in each case. But they also shared a number of more basic, less exclusive, qualities, particularly training, good leadership and discipline.

The purpose of a retreat is to survive so as to fight another day. But both the Light Division and the 1st Marine Division inflicted greater casualties upon their attackers than they themselves sustained. This is not the usual outcome of a withdrawal in the face of a determined and skilful enemy. In both cases, it betokened a self-confident optimism: a never-say-die attitude; an unwillingness to give in and a belief that it's not over until it's over. We will encounter these sanguine qualities again in the next chapter, particularly among the excellent surgeons.

One of Wellington's great strengths as a general was his attention to the details.[8] The successful outcome of both retreats depended on paying close attention to the details as well as to the broader picture.

In the case of the Light Division, the complex and detailed manoeuvres that ensured their safety had been so drilled and practised that they were almost reflexive, even though their success depended crucially upon the timing of their execution.

All battles contain a large element of chaos, but the battlefields of the early nineteenth century were somewhat more predictable than the conditions encountered by the 1st Marines around the Chosin Reservoir. Not only were they facing an entirely new and largely unknown enemy employing unfamiliar tactics, they were also functioning in a relatively alien mode. The primary task of Marines is to storm beaches and hold the ground. But Chosin Reservoir was some 80 frozen miles from their landing point. It is to their enormous credit, therefore, that the Marine leadership came up with a plan of action that embraced the details of surviving the cold, how to beat the Chinese, and how to make a fighting retreat. Successfully coping with the unexpected is the essence of a truly resilient organisation, as we shall discuss in the concluding section of this book.

8 At Apsley House, Wellington's London residence ('No. 1 London'), there is a copy of an order that he wrote in pencil on his knee while he was directing the Battle of Waterloo in 1815. In addition to giving orders for the movements and dispositions of his regiments and batteries, he also gives detailed instructions on how to put out the fire that has begun in the roof of the Hougoumont farmhouse, a critical defensive point in his line.

Chapter 9

Sheer Unadulterated Professionalism

Let me begin by teasing out what I understand by 'professionalism'. It is a fuzzy-edged term embracing a number of different qualities. Aviators call it airmanship; mariners call it seamanship. These labels describe abilities that go well beyond the competent deployment of technical skills. They imply a capacity to see the broader picture, to think ahead and to draw upon a wide range of knowledge and experience so as to perform demanding work safely, elegantly and effectively. It means having a deep understanding of all the various factors that can impact upon task performance for good or ill. It also entails a willingness to engage in all aspects of the job – tedious or otherwise – to the best of one's ability.

When I was writing references for university colleagues, describing a person as a 'true professional' was the highest form of praise in my lexicon. Most full-time academics have three work strands: teaching, research and administration. Not all of these things are equally appealing to all people; but they all need doing. In my view, someone who cheerfully gave each of these activities their best shot displayed professionalism.

So far, I have been considering superior performance under normal conditions. I have reserved the phrase 'sheer unadulterated professionalism' (SUP) for circumstances in which the qualities described above are combined with skill, experience, courage, a cool head and 'grace under fire'. In short, it involves coping with crises and emergencies. All of the heroic recoveries discussed in this chapter involved SUP. In many, but not all, the professionals were as much at risk as those in their care.

Unlike those in Chapter 8, the cases considered here cover a range of domains: the rescue of the *Titanic* survivors; bringing the stricken *Apollo 13* spacecraft safely back to earth – a combined effort involving the astronauts and the controllers at NASA's Manned Space Flight Center in Houston; and two cases in which pilots saved passenger aircraft from disaster – there have been many more such heroic recoveries in aviation history, but these two instances should be sufficient to gauge the mettle of the breed – for now, at least. Finally, we will look at excellent paediatric cardiothoracic surgeons compensating for adverse events during the very challenging 'arterial switch' operation.

Captain Rostron and the Rescue of the *Titanic* Survivors (1912)

Arthur Henry Rostron, Captain of the RMS *Carpathia,* was born in 1869 and first went to sea at 13 as a cadet at the naval training ship, HMS *Conway.* He spent ten years under sail and then joined the Cunard Line where he remained for the next 17 years (aside from a spell in Royal Naval Reserve during the Russo-Japanese War). He began his Cunard career as fourth officer on an ocean liner. He steadily rose through the ranks until he was given his first command in 1907. He was made captain of the *Carpathia* on 18 January 1912 at the age of 43. This was his sixth command in five years.

Captain Rostron was a respected and experienced shipmaster, though he had never responded to a distress signal before. He was known for his rapid decisions and incisive orders. His Cunard shipmates had nicknamed him 'the Electric Spark' for his ability to transmit his own boundless energy to those serving under him. He did not smoke or drink, never used profanity, and frequently turned to prayer. When he prayed, he would lift his uniform cap a small distance off his head, and his lips could be seen moving in silent supplication. There had been little need for quick decisions or prayer in the three days since the *Carpathia* had left New York. But the events of the night of the 14/15 of April 1912 would make a heavy demand on both.[1]

---

1 The material for this section on Captain Rostron came from a number of sources. The most extensive was the excellent book by Robin Gardner and Dan Van der Vat, *The Riddle of the* Titanic (London: Orion, 1995). Other sources came from googling 'Captain Rostron Carpathia'. Among the most informative of these was: 'Arthur Henry Rostron (http://www.geocities.com/hollywood/theater/7937/rostron.html?200811); 'Arthur Rostron', Wikipedia (http://en.wikipedia.org/wiki/Arthur_Rostron); 'The Wireless Operators, the Distress

Before describing the rescue, here is a brief description of the *Carpathia* and her passengers. This twin-screw steam ship had been built by Swan and Hunter in Newcastle, and was launched in August 1902. She had a gross tonnage of 13,603 and was 558 feet in length. Her eight-cylinder quadruple-expansion engines gave her a top speed of around 14 knots. On this voyage, she was carrying 120 first-class and 65 second-class passengers – mostly American tourists bound for a Mediterranean cruise – as well 565 passengers in third class, largely immigrants to the United States returning for a visit to their homelands. Providentially, these numbers meant that her extensive passenger accommodation was only half full. Luckily for the *Titanic* survivors, emigrant ships – the *Carpathia's* main business – tended to be less busy eastbound than westbound.

Shortly before midnight on 14 April, the *Titanic* struck an iceberg a glancing blow and tore open her hull. She began to sink almost immediately. At 2.20 am, the lights went out, and, shortly after, the great vessel nose-dived to the bottom of the ocean. More than 1,500 lives were lost.

Steaming Towards the Location of the Titanic's *Sinking*

Captain Rostron first heard of the impending disaster at 12.35 am, when the 21-year-old wireless operator, Harold Cottam, and the First Officer, burst into his stateroom and woke him up. Cottam had made contact with the *Titanic's* wireless operator almost accidentally. After a long day, he was just about to turn in when, on a whim, he idly scanned the airwaves and tuned into Cape Cod and overheard transmissions to the *Titanic*. He thought he would check to see if the *Titanic* was receiving them. Addressing his transmission to MGY, the *Titanic's* call sign, he tapped out the following message.

> 'I say, OM [old man], do you know there is a batch of messages coming through for you from MCC (Cape Cod) ...'[2]

Call and the Rescue Ship Carpathia' (http://www.titanic-whitestarships.com/Carpathia%20Rescue.htm); and 'RMS Carpathia, 1903–1918' (http://www.sorbie.net/carpathia.htm).

2 The wireless operators were employed by Marconi rather than by the shipping company, and these 'old man' appellations were evidently a part of the house style – though rather incongruous under the circumstances.

At this point, the *Titanic* operator cut across his transmission with the following:

> 'Come at once. We have struck an iceberg. It's CQD [emergency], OM. Position 41 degr. 46 min. N, 50 degr. 14 min. W.'

The shocked Cottam asked 'Shall I tell my captain? Do you require assistance?' The reply to these superfluous questions was 'Yes. Come quick.'

From that moment on, Captain Rostron's decision-making talents went into overdrive. He immediately ordered the *Carpathia* to be turned around, and only then asked Cottam if he was sure – most captains would have done it the other way round. He then told Cottam to inform the *Titanic* that the *Carpathia* was coming as fast as it could. They should expect them to be there within four hours. The *Titanic* operator replied 'TU OM (Thank you, old man)'.

Still half-dressed, Rostron rushed to the chart room. He glanced down the meridians and picked off the degrees and minutes of *Titanic's* position. Then he calculated his present position and worked out the course to the *Titanic*. On the bridge, he instructed the helmsman to steer 'North 52 West'. Later he testified that he did all of this while he was still dressing.

Meanwhile the other officers, including the chief engineer, had assembled on the bridge. The captain led them into the chart room and quickly explained the situation. Nowhere is Captain Rostron's consummate professionalism more apparent than in the string of orders he subsequently issued. He thought of everything down to the smallest detail. Here is a flavour of his instructions:

- He told the First Officer to cease all routine work in order to organise the ship for rescue operations. All seamen were to be on deck to keep a sharp lookout and to swing out the boats. Electric light clusters were to be rigged at each gangway and over the side.
- He ordered the chief steward to call out every man and have them prepare coffee for all hands. They were also to have hot soup, coffee, tea, brandy and whisky ready for the survivors. Blankets were to be piled at every gangway. Smoking rooms, lounge and library were to be converted into dormitories for the

rescued group. All of the *Carpathia's* steerage passengers were to be grouped together, and the space saved used for the *Titanic's* steerage passengers.

- He asked the purser, assistant purser and chief steward to receive the rescued survivors at different gangways and assist them to the dining rooms and other quarters adapted for their reception. A purser was to be stationed at each gangway to get the full names as soon as possible in order that a list of the saved should be sent by wireless. He also ordered that all officers' cabins should be given up for the survivors.
- He called the ship's surgeon, Dr McGhee, and asked him to collect all the restoratives and stimulants on the ship and to set up first aid stations in each dining saloon. He also instructed that the two other doctors on board, an Italian and a Hungarian, should be in charge of the second- and third-class saloons respectively. Dr McGhee was to manage the first-class saloon.
- Captain Rostron further instructed that all gangway doors were to be opened and bosun's chairs to be slung out at each gangway. Pilot ladders were to be dropped over the side. Portable lights and nets were to be draped along the side of the ship to make boarding easier. Oil was to be got ready to quieten the sea should it be too rough in the vicinity of the life boats. He warned them that they might have to pick up more than 2,000 people. Sadly, the reality proved to be less than half that number.
- As he gave orders, the captain urged them all to be as quiet as possible. The task ahead would be tough enough without having the *Carpathia's* passengers underfoot. He instructed that stewards should be posted in every corridor who were to tell any wandering passengers that the ship was not in trouble and to persuade them back to their cabins. He also sent the master-at-arms and a special detail of stewards to keep the steerage passengers under control. No one knew how they would respond to being shuffled around.
- But his bravest act was to tell the chief engineer to pile on as much coal as possible and to call out the off-duty watch to assist with the stoking. All the heat and hot water to the cabins should be cut off so that as much steam as possible could be piled into the boilers. Presently, the ship vibrated as the engine revolutions steadily increased. Within a few minutes of the *Titanic's* call for

assistance, the old *Carpathia* was slicing through water at 17 knots, a speed that was three knots faster than the theoretical top speed of 14 knots. The ship maintained this rate of knots for some three hours until they reached the edge of the icefield.

- Rostron called the Second Officer over to the starboard wing of the bridge and told him 'Station yourself here, Mister, and keep a special lookout for lights or flares – and for ice! In this smooth sea it's no use looking for the white surf around the base of the bergs, but you will look for the reflection of starshine in the ice pinnacles. We'll be into the icefield at 3 am, or perhaps earlier. Extra lookouts will be posted on the bows and in the crow's nest, and on the port wing of the bridge, but I count on you with your good eyesight, and with God's help, to sight anything in time for us to clear. Give that all your attention.' The Second Officer remembered glancing over at the bridge where the captain was standing alone with his head bowed and his lips moving in prayer. The captain was to say later that it was God's hand on the helm rather than his.

At about 2.45 am, the Second Officer called out that there was an iceberg glimmering in the starlight about three-quarters of a mile ahead. The captain immediately altered course to starboard and reduced to half speed. From the port wing of the bridge he could see that they had cleared the iceberg and no more ice was in sight. He went back to the telegraph and signalled 'Full Speed Ahead'.

A few minutes later, they spotted another iceberg, and then a whole succession of them. For half an hour or so, the *Carpathia,* steaming at its forced top speed, zigzagging among the mountainous ice islands, avoiding each one in turn. For Rostron, reducing speed was out of the question: time was of the essence. He took a calculated risk.

At the subsequent inquiry in New York, Senator Smith pressed Captain Rostron on the legitimacy of his decision to cut through the ice field at full speed. He replied:

> 'I can confess this much, that if I had known at the time there was so much ice about, I should not [have steamed at such a speed]; but I was right in it [the ice field] then. I could see the ice. I knew I was perfectly clear. Although I was running the risk with my own ship and my passengers, I also had to consider what I was going for.'

'To save the lives of others,' the senator prompted. 'Yes, I had to consider the lives of others,' Rostron replied. The senator's eyes filled with tears and he was unable to speak. Mopping his tears with a handkerchief, Senator Smith choked out 'I think I may say for myself that your conduct deserves the highest praise.' Captain Rostron, similarly overcome, said quietly 'I thank you, sir.'

Cottam, the young wireless operator, received the last message from the *Titanic* at just before 2 am. It said: 'Engine room full up to the boilers.' The *Titanic* had a little over half an hour to go before sinking.

Meanwhile the *Carpathia* raced on, giving out a huge plume of black smoke, weaving between icebergs and firing rockets over her bow – one every 15 minutes, with Cunard Roman candles in between. Captain Rostron wanted to reassure the survivors that help was on its way.

The Rescue

They were now very close to the position given in the distress call – Lat 41 46 N. Long 50 14 W. But by 3.35 am there was still no sign of the lifeboats or the ship – the *Carpathia* crew knew that the *Titanic* was sinking, but were not yet aware that it had already sunk an hour earlier. At 3.50 am, Rostron put the engines on 'stand by' and at 4 am, he stopped. The first light of day was just appearing.

Just then a green flare was seen, dead ahead and low in the water. Its light revealed the outline of a lifeboat around 300 feet away. A swell had got up, and the boat could be seen rising and falling but it was hardly moving forwards, as if those manning the oars were exhausted. Rostron started up his engines and began to manoeuvre the *Carpathia* to starboard, putting the lifeboat on his leeward side. Owing to an iceberg that lay directly ahead, Rostron had to change direction so that he had to pick up the survivors on his starboard side. He sent two quartermasters over the side to fend the lifeboat off so that it didn't bump the ship's side.

The first to be rescued was the No. 2 lifeboat with Fourth Officer Boxhall in charge and 24 other occupants aboard. On the bridge, Rostron knew without asking that the *Titanic* had sunk, but he felt he had to go through with the formalities. He asked Boxhall to come to the bridge. He stood before the captain shivering and suffering from shock. 'The *Titanic* has gone down?' Rostron asked. 'Yes,' said Boxhall, his voice breaking, 'she went down at about 2.30'. Rostron asked if there were many people on board when she sank. 'Hundreds and hundreds, perhaps a thousand, perhaps more,' cried the distraught Boxhall. Rostron sent him below to drink some hot coffee and get warm.

Dawn revealed well over 20 large icebergs and scores of lesser bergs. They dwarfed the lifeboats spread over a four-mile area. Rostron described the scene as follows:

> Except for the boats beside the ship and the icebergs, the sea was strangely empty. Hardly a bit of wreckage floated – just a deckchair or two, a few lifebelts, a good deal of cork; no more flotsam than one can often see on a seashore drifted up by the tide. The ship had plunged at the last, taking everything with her. I saw only one body in the water; the intense cold made it hopeless for anyone to live long in it.

The business of manoeuvring a large ship among the ice floes without damaging the small lifeboats was extremely challenging and taxed Captain Rostron's seamanship to the limit. The difficult task of recovering the survivors took hours. It was not until 8.30 am that the last lifeboat was emptied. By then, 705 of the *Titanic*'s survivors had been rescued.

Rostron now wondered where he should take his unexpected guests. Halifax was nearest, but there was ice along the way. The Azores was best for the *Carpathia's* schedule, but he did not have the linen or provisions to last that far. The costliest option for Cunard Line was New York, but it was best for the passengers. Rostron set sail for New York, though they had to spend four hours skirting around the ice field.

En route to New York, the Captain put strict limits on communications with the shore in order to minimise false hopes and false information. He also imposed a news freeze. He was saving his communication equipment for official traffic and private messages from the survivors.

Arriving in New York and the Aftermath

The *Carpathia* entered New York Harbour in the evening of 18 April. There, the weather changed completely as a thunder storm hit the ship. Rostron insisted on landing the survivors as quickly as possible, and the first of them came off at 9.35 pm. Nor did he allow the rescued passengers to be harassed by the horde of news reporters present on the quay. One managed to scramble onto the ship, but he was told that under no circumstances would he be allowed to interview the survivors on board. Rostron left the reporter on the bridge on his own honour. Later Rostron said of him: 'He was a gentleman.'

Captain Rostron was justly rewarded for his extraordinary achievements. He was awarded the Congressional Gold Medal by President Taft and the American Cross of Honor, among several other medals and awards. After the First World War, he was knighted by King George V. He was made commodore of the Cunard fleet before retiring in 1931. He died of pneumonia on 4 November 1940 and is buried at West End Church in Southampton. His wife died three years later and was laid beside him.

One final postscript: on 17 July 1918, the *Carpathia* was travelling in a convoy bound for Boston when she was struck by two torpedoes some 120 miles off Fastnet, southern Ireland. A third torpedo struck the ship as the lifeboats were being manned. Five of the crew were killed by the explosions. The remainder of the crew and 57 passengers were picked up by HMS *Snowdrop* and brought safely to Liverpool.

Saving *Apollo 13* (1970)

This event has a special significance for me because I was a part-time contractor for NASA at the time and was visiting NASA Ames in Mountain View, California, in the summer of 1969 when the director called us all into the conference room to hear early news of the outcome of the inquiry into the near-disaster.

'Houston, we've had a problem.' These must be among the most oft-quoted words in the late twentieth century after 'One small step for man ...' In reality, having a liquid oxygen tank explode and blow out the side of *Apollo 13*'s service module was just the beginning of a succession of problems. Life has sometimes been defined as 'one damned thing after another'. That's a fair description of what happened in recovering *Apollo 13*. The saving of *Apollo 13* took several days and was extremely technical, so I will simplify the story by organising it around the succession of problems – and problems solved – that brought the three astronauts safely back to Earth.[3] But first I must say something about the spacecraft and its crew.

The Mission, the Spacecraft and the Astronauts

Apollo 13 was to be the third manned lunar landing. It was launched on 11 April, 1970, at 13.13 Houston time (for the superstitious, there were some ominous numbers here – and the explosion that blew out the side of the service module occurred on 13 April). Each mission was lifted off from Cape Canaveral, Florida, by an 85-metre tall, three-stage Saturn V booster. At the top of the launch rocket were two spacecraft: a three-person mother ship to go to the moon and back (the command and service module or CSM), and a two-person lunar module (LM) designed to travel between the CSM and the surface of the moon. Each of the two spacecraft was divided into two parts. The CSM comprised a cylindrical service module with a conical command module on top. The service module housed the main engine and provided all the oxygen, electricity and water needed for the six-day round trip between the Earth and the moon. The command module housed the crew, the flight computer and the navigation equipment. It was the only part of the Apollo stack that was designed to return to Earth, plummeting through the atmosphere with its blunt end downwards.

The lunar module was made up of an ascent stage and a descent stage. The latter had a powerful engine used to land the LM on the moon. After the lunar excursion was complete, it served as a launch pad for the ascent stage, which carried the astronauts back to the command module. After blast off from the moon, the ascent part of the LM rendezvoused with the CSM in lunar orbit. On this mission, the CSM was called *Odyssey* and the LM, *Aquarius*. En route to the moon, the *Odyssey* and the *Aquarius* were docked nose to nose. The astronauts remained in the command module, and the electrics in the LM were switched off to preserve power.

3 The main sources for this account of the recovery of *Apollo 13* are: Cass, S. (2005) *Apollo 13, We Have a Solution* (IEEE Spectrum: http://www.spectrum.ieee.org). Compton, W.D. (2000) *The Flight of Apollo 13 (excerpts)* (http://liftoff.mfsc.nasa.gov/Academy/History/APOLLO-13/compton.html).

The mission commander on *Apollo 13* was James A. Lovell. The command module pilot was John L. Swigert, who was a fairly late replacement for Thomas K. Mattingley who may have contracted German measles. The third member of the crew was Fred W. Haise Jr, the lunar module pilot.

The Initiating Problem

Most of the power on board came from three fuel cells in the service module. Figure 9.1 shows a cutaway diagram of the contents of the service module. The fuel cells (at the top) provided water and electricity by combining oxygen and hydrogen stored in cryogenic tanks below. The two oxygen tanks were approximately round in shape, and the hydrogen tanks, below them, were more cylindrical. It was the nearest oxygen tank (Number 2) that exploded at 9 pm on 13 April some 56 hours into the flight. The astronauts were lucky to escape with their lives.

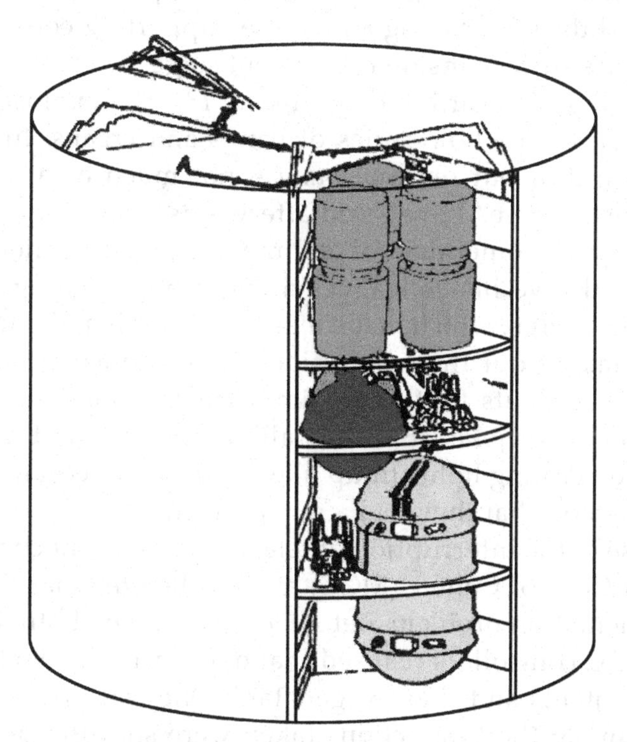

Figure 9.1 **A cutaway diagram of the service module. It was the round cryogenic oxygen tank (second layer) that exploded and nearly destroyed the spacecraft**

The oxygen was in liquid form and needed stirring in order to obtain accurate quantity readings. This was done by a paddle wheel inside the tank and was activated by a switch in the command module. Due to a series of mistakes and misadventures during pre-launch testing, the Teflon insulation had been burned away. The tank survived the first two activations of the propeller (one every 24 hours), but Tank Two's quantity sensor failed on the second cryo-stir. Mission control asked the astronauts to stir the cryos every six hours to help establish how much oxygen remained in Tank Two. There were five applications of current to the bare wires. It was the last of these that caused the explosion.

In some ways the timing was providential.[4] Had the explosion occurred later, it might have been after the lunar lander had departed for the moon. In which case, the LM would not have been available to act as an indispensable 'life boat' and there could have been no recovery. Or it might have happened after the LM had rejoined the CM leaving all the life-supporting consumables on the moon's surface inside the descent stage.

In the Ron Howard movie *Apollo 13*, the explosion was accompanied by a whole series of bangs and creaks. In reality, 'there was a dull but definite bang – not much of a vibration though – just a noise,' Lovell said afterwards.

Fred Haise, something of a prankster, had earlier found that he could give his colleagues a fright by actuating a lunar module relief valve, which made a loud bang. When the fuel tank exploded, taking out the whole side of the service module, the other two astronauts first gave him accusing looks, thinking it was another of his jokes. But then all of the command modules' caution and warning lights lit up like Christmas decorations, and they understood that they had a real problem

Because of the interruption of telemetry, the flight controllers at Mission Control Center (MCC) at first thought that they had an instrumentation problem. Before long, however, both the crew and the MCC controllers realised that oxygen Tank Two had lost all of its contents and that oxygen Tank One was also losing its oxygen. Among the first actions taken were shutting down one

4 Woodfill, J. *The Stir that Saved the Lives of* Apollo 13*'s Crew* (http://www.dunamai.com/articles/American_History/stir_saved_Apollo_13.htm).

fuel cell and switching off non-essential systems in the CM in order to minimise power loss. Shortly after, the second cell was shut down as well. When the remaining oxygen ran out, the CM would only have three re-entry batteries providing 120 ampere-hours. But these were essential in the final stages of the mission. Gene Kranz, the flight director, decided to use the LM as a lifeboat, and ordered his team to provide minimum power in the LM to sustain life. But that was not as easy as it sounds.

Powering Up the Lunar Module

Although the LM had big, charged batteries and full oxygen tanks designed to last for the 33 hours of the lunar excursion, almost everything on *Aquarius* was switched off aside from some heaters necessary to prevent critical systems from freezing. They were powered by two umbilicals from the command module, which in turn got its power from the service module.

Within the *Odyssey*, the umbilicals were connected to a power distribution switch that shifted the LM between drawing power from the *Odyssey* and drawing power from its batteries. But here was the problem: the distribution switch itself needed electricity to operate, which the *Odyssey* could no longer supply.

This is where professionalism saved the day. A year earlier, the simulator people had failed the CM's fuels cells at almost the same point into the mission as the *Apollo 13*'s explosion had occurred. Bob Legler, a lunar module flight controller, had been present on that occasion when the LM was suddenly in demand as a lifeboat. Some lifeboat procedures had already been worked out, but none involved having a damaged CM attached. The simulation finished with a dead crew – a virtual crew, but dead nonetheless.

In the subsequent months, the lunar module controllers worried away at the problem, examining many different failure scenarios and coming up with solutions for each. They quickly pulled them off the shelf after the *Apollo 13* explosions. The time gained in this way stood them in good stead. When the LM was powered up sufficiently to sustain life, there were just 15 minutes of life left in the last fuel cell aboard the *Odyssey*.

Programming the Lunar Module's Guidance System

After the explosion, the crucial issue was how to get the crew back to Earth as quickly as possible. The flight dynamics team at MCC calculated that if the astronauts used the *Odyssey's* engine and burned every last drop of fuel, they could turn around and come straight back to Earth. But the main engine was in the service module and no one knew how far it might have been damaged by the explosion. It might malfunction, or it could explode, killing the crew instantly.

The other option was to let *Apollo 13* continue to the moon. There, gravity would pull the spacecraft around the back of the moon and hurl it back toward the Earth. But the return journey would take several days and the LM was designed to support two astronauts for two days – not three men for four days.

After much agonising, Gene Kranz elected for the moon option. Kranz said this was his toughest call. But he had great faith in the lunar module and the LM controllers. 'I was pretty much betting that this control team could pull me out of the woods once we had decided to go around the moon.' In the film, his subsequent announcement ended with: 'Failure is not an option.' Kranz claimed afterwards that he never actually uttered these words; but he liked them so much they became the title of his book.[5]

There was yet another problem. For various operational reasons, *Apollo 13* was not on a free-return trajectory. If nothing was done to correct the trajectory, the spacecraft would head back towards the Earth, but miss it by several thousand miles. In order to obtain a freereturn, on target trajectory, they needed to use the *Odyssey's* guidance system. But this was usually powered off for most of the way to the moon, as it was on this occasion. So it didn't know where it was.

The LM module's guidance system had a computer identical to the one in the command module. This meant that the crew would have to transfer the alignment data from the command module's computer to the equivalent one in the LM. But because the LM and CM were docked head to head, the angles would have to be inverted, requiring some old-fashioned arithmetic. The job

5 Kranz, G. (2000) *Failure is Not an Option*. New York: Simon & Schuster.

fell to James Lovell, but because he had made many arithmetical errors during simulation, he asked the controllers to check his numbers.

As soon as this was completed, they pulled the plug on the command module, shutting it down completely. But it would have to be powered up again before re-entry. And that posed another problem since the CM was never designed to be switched off in flight.

The Carbon Dioxide Problem

As *Apollo 13* sped towards Earth, mission control was wrestling with yet another problem. While the LM had enough oxygen to accommodate Swigert as well as the intended LM crew of Lovell and Haise, carbon dioxide was beginning to build up because the lithium hydroxide canisters (designed to absorb carbon dioxide) were being overwhelmed. The *Odyssey* had enough spare canisters on board but they would not fit into the holes intended for the LM's round canisters.

Ed Smylie, a life-support systems engineer, had anticipated the problem. For two days he and his team worked on how to jury-rig the *Odyssey*'s square canisters so as to marry up with the LM's round holes. Now using materials available on board – a sock, a plastic bag, the cover of a flight manual and lots of duct tape – the crew put together Smylie's strange contraption and taped it into place. It worked and the carbon dioxide levels fell back into the safe range. Mission control had delivered another miracle.

Realigning the Trajectory

Although the two previous burns (rocket engine firings) had been spot on, the flight controllers were worried by the fact that something was pushing the spacecraft off course – later it turned out to be a water vent on the *Aquarius* acting like a small rocket jet. They needed another burn to correct the trajectory. But the navigation system was non-operative. They needed some way to align the spacecraft properly for the corrective burn.

Chuck Dietrich, one of the flight controllers, recalled an alignment technique that had been used during the *Mercury*,

Gemini and *Apollo* Earth-orbit missions. It involved pointing the spacecraft at a reference point on the Earth's surface, in this case the terminator between night and day. But this technique had never been used on a return from the moon. If *Apollo 13* missed the entry corridor by only a single degree, the spacecraft could skip back into space or burn up in the atmosphere.

The crew was cold and very tired. They had got very little sleep since the explosion and the temperatures on board had dropped almost to freezing. But they performed the course-correction manoeuvre – and a second one a day later – perfectly.

Powering Up the Command Module for Re-entry

Three days after the explosion and it was time to prepare the spacecraft for re-entry. The first step was to recharge the batteries in the command module. When putting the LM into lifeboat mode, Legler and his team had worked out a way to run power from the LM to the CM along the umbilicals that connected the spacecraft. The same procedure could be used to recharge the *Odyssey's* batteries.

Even with fully charged batteries, the *Odyssey* risked running out of electricity before it splashed down. Once a system had been turned on in the CM, it had to stay that way. The main problem was to determine the smallest number of systems that could be turned on, and how long to wait before activating them.

John Aaron – the 'go-to' person in MCC when things went wrong – worked out an inspired procedure for performing the power-up procedure. Normally one of the first things to be turned on is the instrumentation system. But for this power-up the instrumentation would be turned on last.

This required the crew, and especially the CM pilot, Swigert, to perform the entire power-up sequence without instrumentation. If they made an error it could be too late to fix when the instrumentation was finally turned on. Despite their fatigue, the astronauts performed perfectly. Chris Kraft, a pioneer mission controller, commented: 'That's why we chose test pilots to be astronauts. They were used to putting their lives on the line, used to making decisions, used to putting themselves in critical situations. You wanted people who would not panic under those circumstances. These three guys [Lovell, Swigert and Haise], having been test pilots, were the personification of that theory'.

Splash Down

Once powered up, the crew moved back into the command module and strapped in. As part of the re-entry procedure, they jettisoned the damaged service module, taking pictures of the huge gash in its side. Then it was time to abandon the lunar module. 'Farewell, *Aquarius*, and we thank you,' said Lovell as he watched the LM drift slowly away.

It was another hour before the command module met the first wisps of the Earth's atmosphere. In few more minutes the astronauts would either be home free or dead. A few seconds later, the *Odyssey* disappeared into dense radio static. The controllers calculated that it would take three minutes for the spacecraft to regain radio contact. But when the appointed time came, they still had received no word of the *Odyssey*. 'It was the worst time of the whole mission,' said Kranz.

Seconds later, the CM's signal was picked up by a circling communications aircraft. The main parachutes still had to be deployed and disaster was still a possibility. Then the three red-and-white canopies were picked up by a TV camera on board the *USS Iwo Jima*, the aircraft carrier leading the recovery effort. Then the three astronauts were seen on the deck. They had made it home.

Postscript

The account I have given of the saving of *Apollo 13* has only mentioned a few names. But hundreds of people were involved in the rescue: off-duty controllers, astronauts and simulation technicians, contractor personnel and many more. It was truly a team effort, and a magnificent display of sheer unadulterated professionalism, both in the spacecraft and on the ground.

British Airways Flight 09 (1982)

Flight 09 was a scheduled British Airways flight from London Heathrow to Auckland, with stops in Bombay, Madras, Kuala Lumpur, Perth and Melbourne. On 24 June 1982, the route was flown by a 747-236B jumbo jet with the name *City of Edinburgh*. On board were 248 passengers and 15 crew members. In the cockpit were Captain Eric Moody, Senior First Officer Roger Greaves and Senior Engineer Officer Barry Townley-Freeman.

The trouble began shortly after 13.40 GMT (20.40 Jakarta time) when the aircraft was flying above the Indian Ocean, south of Java. The captain had stepped out for a toilet stop and the two remaining flight crew noticed an effect on the windscreen similar to St Elmo's fire. It was as if it was being hit by tracer bullets. This persisted when the captain returned to the flight deck. Despite seeing nothing on the weather radar, they switched on the engine anti-ice and the passenger seat belt signs as a precaution.

Meanwhile, the passenger cabin started to fill with smoke. As it got thicker passengers became increasingly alarmed, particularly when those by the windows announced that the engines were unusually bright. It was as if they had a powerful light in them and was shining out through fan blades. A curious strobe effect made the engines appear to be rotating slowly backwards.

At approximately 13.42 GMT (20.43 Jakarta time), the number four engine surged and then flamed out. The First Officer and the flight engineer immediately performed the engine shutdown drill, closing off the fuel supply and arming the fire extinguishers. The captain added some rudder to counter the uneven thrust.

At this point the passengers spotted long yellow glows coming out of the remaining engines. Less than a minute after the number four engine failed, the number two engine surged and flamed out. Engines one and three shut down almost simultaneously. The flight engineer announced 'I don't believe it – all four engines have failed.'[6] The 747 had become a glider.

Most of the engine instruments were useless, either having frozen or fallen to zero. One airspeed indicator showed 270 knots, and the other one 320 knots. Unknown to them, the aircraft had entered a large cloud of volcanic dust caused by an eruption of Mount Galunggung, some 110 miles south-east of Jakarta. This did not show up on the weather radar because the dust was dry.

A 747 aircraft can glide 15 kilometres for every kilometre it loses in height. Captain Moody estimated that from its flight

6 British Airways Flight 9, Wikipedia (http://en.wikipedia/wiki/British Airways_Flight_9).

level of 11,280 metres (37,000 feet), the 747 would be able to glide for 23 minutes and cover 261 kilometres (141 nautical miles). At 13.44 GMT (20.44 Jakarta time), the captain told SFO Greaves to declare an emergency to Jakarta Area Control, telling them that all four engines had failed. But the air traffic controllers misheard the message and believed that only engine number four had shut down. It was only after a Garuda Indonesia flight relayed the correct message to them, that they understood the true nature of Flight 09's plight.

The total engine shutdown was immediately obvious to the passengers. Some wrote notes for their loved ones, such as Charles Capewell's scribbled message on the cover of his ticket: 'Ma. In trouble. Plane going down. Will do best for boys. We love you. Sorry. Pa XXX.' Others screamed that they were going to die, and still others attempted to calm down the panicky ones.

In the cockpit, the crew tried to contact Jakarta for radar assistance. But they could not be seen by Jakarta radar, perhaps because of high mountains in the vicinity. An altitude of at least 11,500 feet was required to cross the coast safely. Captain Moody decided that if the aircraft was unable to maintain altitude by the time they reached 12,000 feet, he would turn back out to sea and attempt to ditch. The crew started engine restart drills, but they were unsuccessful.

At this point, Captain Moody made an announcement to the passengers over the PA system. It has been called 'a masterpiece of understatement':

> Ladies and Gentlemen, this is your Captain speaking. We have a small problem. All four engines have stopped. We are doing our damnedest to get them going again. I trust you're not in too much distress.

At 13,500 feet, the flight crew tried one last engine restart procedure before turning towards the ocean and the very risky business of ditching on water. Although there was a procedure for it, no one – before or since – had ever tried it in a 747. Then they finally struck lucky. Number four engine started and the Captain used its thrust to reduce the rate of descent. Shortly after, the remaining three engines restarted, thanks to having one generator still operating, allowing engine ignition. The crew requested an increase in altitude to 15,000 to clear the mountains. But as the aircraft approached its target altitude, the St Elmo's fire started hitting the windscreen. The captain throttled back but this did not prevent the number two engine surging again and it had to be shut down. They immediately descended to 12,000 feet.

As they approached Jakarta, they were told that it was a clear night and there was good visibility at the airfield. Unfortunately, that wasn't true for the aircraft. It was very difficult to see anything through the windscreen and they had to make the approach almost entirely on instruments and without the Instrument Landing System. Captain Moody described the experience 'as a bit like negotiating one's way up a badger's arse'. The runway lights could be seen through a small strip of the windscreen, but the aircraft's landing lights were not working. Having landed, they found that they could not taxi because the glare from the apron lights made the windscreen opaque, so they to be towed in by a tug.

The *City of Edinburgh* (dubbed 'the flying ashtray') entered the *Guinness Book of Records* as having flown the longest glide in a non-purpose-built aircraft.

The crew received various awards, including the Queen's Commendation for Valuable Service in the Air and medals from the British Air Line Pilots Association (BALPA). BALPA also included the crew in their Outstanding Airmanship Award booklet.[7] Other inclusions in this booklet will be discussed in this and later chapters.

The BAC 1-11 incident (1990)

The term 'incident' is something of an understatement, as we shall see.[8] Flight BA 5390 was a charter flight en route to Malaga. It left Birmingham International Airport at 0720 on Sunday 10 June, 1990. The weather was fine and the flight was anticipated to take three hours. But they never left UK airspace.

---

7 The British Airline Pilots Association. *Outstanding Airmanship Award*. Harlington: BALPA.

8 'Oh, what a flight – the story of BA 53900'. (http://www.bbc.co.uk/dna/h2g2/A20460782). Frow, I. *BAC 1-11. Outstanding Airmanship Award*. Harlington: The British Airline Pilots Association (www.balpa.org).

On the night of 10/11 June, the aircraft was scheduled to have a window change. The shift manager elected to do the job himself because the shift was short-handed. I was involved in the subsequent investigation and have elsewhere written an account of the defensive failures and errors that led to the window being secured by bolts that were fractionally thinner than those required for the job,[9] and so I will focus here upon what happened during the subsequent flight.

The pilots were Captain Tim Lancaster and Senior First Officer (SFO) Alistair Atcheson. During their pre-flight check of the maintenance log, they noted that the left-hand (captain's) window had been changed. But there was nothing to suggest that this would be anything but an ordinary flight.

In addition to the two flight crew, there were four cabin crew and 81 passengers on board. Captain Lancaster had flown with the cabin crew on a number of previous occasions and was familiar with their routine. It was the co-pilot's first outing with this group.

Thirteen minutes into the flight, the aircraft was climbing through a height of 17,000 ft and was over Didcot in Oxfordshire. The cabin crew were preparing breakfast for the passengers and the pilots in the galley, just the other side of the flight deck door. Anticipating the arrival of his breakfast, Captain Lancaster undid his shoulder harness and lap strap and engaged the autopilot, handing over control to SFO Atcheson. Then the captain watched in horror as his windscreen began to move slowly outwards. Suddenly, there was a loud bang and the windscreen shot out of the aircraft. The air misted over due to the explosive decompression and the flight deck door came off its hinges and lay across the radio and navigation console. Even worse, Captain Lancaster followed the window partially out of the aircraft, and was pinned back against the roof of the cockpit. His shirt had been ripped off his back and his legs had become trapped around the control column pushing it forward. He had also knocked the autopilot and VHF switches to 'off' in the process. The plane began to dive and roll.

---

9 Reason, J., and Hobbs, A. (2003) *Managing Maintenance Error: A Practical Guide*.Aldershot: Ashgate, (p. 41).

The first member of the cabin crew to enter the devastated flight deck was Nigel Ogden. He had been preparing the captain's breakfast in the galley on the other side of the now absent door. He acted quickly and was able to grab Tim Lancaster around the waist in order to drag him back inside. But the captain's legs had become wedged between cockpit coming and the control column, limiting control movement. In addition, Nigel Ogden and the remains of the cockpit door were also lying on the control column, jamming the throttle to climb power and covering the speed brake lever. A 400 mph wind was blowing through the aircraft, but Nigel Ogden persisted in his efforts to save the captain and the plane by trying to pull him back inside the cockpit and release his feet from the control column.

Meanwhile, Alistair Atcheson had only released his shoulder harness and was still in his seat trying to re-engage it. He was faced with a decompression, a captain jammed halfway out of the window and an aircraft that was banked 25 degrees in a spiral dive. The throttle thrust levers were jammed open and there was only partial control movement. As soon as he was secure, he put on his oxygen mask and began to fight with his control column in an effort to pull the aircraft out of its uncontrolled dive. At the same time he was also trying to contact London Control with a Mayday message. Air traffic control heard the Mayday call, but were not able to determine what the problem was.

The dive continued and the aircraft plunged through some of the busiest air space in Europe, risking collision with any of several other aircraft in the vicinity. Eventually, SFO Atcheson managed to level the plane at 11,000 feet and reduce the speed to 180 mph. At that height, there was no need for oxygen. This allowed him to reinstate the autopilot and establish two-way communication with the control tower.

At this point, John Heward, the chief steward, removed the flight deck door and stored its remains in the nearest toilet. Then he strapped himself into the left-hand observation seat and held on the captain's feet to assist Nigel Ogden who, though suffering the effects of exposure and frostbite, was still clinging on.

The speed reduction caused the captain's rigid body to slide to the side or the aircraft where he could be seen by those on the flight deck. They could see that his eyes were open, but there was no sign of life. The three men in the cockpit exchanged a glance, and the co-pilot shook his head to mean that the stewards should not let the captain go even though he appeared to be dead. If they let him go there was real danger that the port wing or engine could be damaged. Besides which, his body was partially blocking the gaping hole where the windscreen had been.

The co-pilot had asked to land at Gatwick as he was familiar with it, but due to the volume of traffic around both Gatwick and Heathrow, he was directed to land at Southampton. Southampton was closer, but all the maps and charts had been lost in the blow-out. Having never landed there, Alistair Atcheson was worried not only about his unfamiliarity with the airfield, but also that the runway was some 400 metres shorter than the 2,200 metres required for landing the BAC 1-11. In addition, the wings were still heavy with fuel. The conditions for the landing were far from ideal – to say the least.

The co-pilot called upon all of his 7,500 hours of flying experience as he began his descent. But they had another surprise in store. When the flight reached 300 feet, Captain Lancaster's legs began to kick, giving some hope that he might still be alive. He was, though suffering a broken right arm and wrist, a fractured left thumb, bruising, frostbite and shock. He returned to work five months later.

The flight landed safely at 07.55 am. The passengers disembarked shocked but unhurt. The fire and ambulance crews released Captain Lancaster through the cockpit. He regained consciousness briefly before being taken to hospital.

For his outstanding courage, piloting skill and airmanship, Alistair Atcheson was awarded the Queen's Commendation for Valuable Services in the Air, the Gold Medal for Airmanship and the British Airways Award for Excellence – among several other honours and medals. But perhaps the most moving tribute to the crew was made when they returned to Birmingham International Airport. The entire concourse fell silent as they walked through, and then erupted with spontaneous applause.

Surgical Excellence (1995–97)

Background

In the mid- to late-1990s, I was fortunate to be part of a research team investigating the impact of human and organisational factors upon the outcome of a high-technology surgical procedure. The team was led by Professor Marc de Leval, a consultant paediatric surgeon at Great Ormond Street Hospital for Sick Children, and the principal human factors specialist was Dr Jane Carthey (a tutee of mine at the University of Manchester during her undergraduate years). The success of this project hinged very largely on her personal and professional skills. Most of what will be discussed below came from her observations of the performance of surgical teams during the long and arduous arterial switch operation (ASO).[10]

Dr Carthey was present at 165 of these procedures (out of the total of 243 performed in the UK over an 18-month period, January 1996 to June 1997). Twenty-one consultant surgeons in 16 institutions throughout the country performed these neonatal switch operations. Of these, 19 surgeons were directly observed. The data reported below derives in part from 16 of these 19 surgeons: inclusion being dependent upon whether the surgeon in question performed sufficient procedures to make compensation comparisons statistically meaningful.

The Arterial Switch Operation

The ASO is carried out on babies (< 35 days) who are born with the great vessels of the heart connected to the wrong ventricle: the aorta is connected to the right ventricle and the pulmonary artery to the left ventricle. This procedure involves correcting these congenital defects by transposing the pulmonary artery and the aorta so as to permit the full circulation of oxygenated blood. Without such an intervention, the child would die. The operation involves putting the patient on to a heart–lung bypass machine and freezing the heart with a potassium-based solution (cardioplegia). The procedure may last for five or six hours and

10 Carthey, J., de Leval, M., Wright, D., Farewell, V., Reason, J., and all UK paediatric cardiac centres (2003) 'Behavioural markers of surgical excellence.' *Safety Science*, 41: 409–425.

is highly demanding both technically and in human terms. For some 90–120 minutes of this time, the child's heart is stopped while the surgeon transects the native aorta, excising the coronary arteries from it, and re-implanting them into a neo-aorta. A neo-pulmonary artery is also reconstructed using the tissue from the trunk of the native aorta and a piece of the pericardium.

The most challenging part of the procedure involves relocating the coronary arteries, each comprising very thin friable tissue. Indeed, the arterial switch procedure takes the surgical team – and particularly the consultant surgeon – close to the edges of the human performance envelope on a variety of parameters: in psychomotor skills and decision-making, and in its claims upon knowledge, experience, leadership, management and communication skills. Errors of one kind or another are almost inevitable under such conditions. What matters are not the errors per se but whether or not they are detected and recovered. In the surgical context, as we shall see, bad outcomes happen when major adverse events, usually the result of errors, go uncompensated; happy outcomes – by far the majority – are due in large part to effective compensation by the surgical team.

The Outcomes of the Procedure

Outcome data were available for 230 patients. The outcome of surgery was assessed on a four-point scale, as shown below. The numbers in parentheses indicate the percentage of patients falling into each outcome category:

- *Level 1*: Child extubated in less than 72 hours (49 per cent)
- *Level 2*: Extubation took longer than 72 hours (26 per cent)
- *Level 3*: Near miss (long-term morbidity) (18.5 per cent)
- *Level 4*: Death (6.5 per cent)

Major and Minor Events

Events were anything untoward that disrupted the intended flow of surgical activities. The large majority of these events arose because of human error. They were of two kinds:

1. *Major events*: These were occurrences that were each severe enough to pose a direct safety threat to the patient. They included:

instability due to rough handling of the blood vessels; delay in administering heparin (an anti-coagulant); decision errors, particularly with regard to dealing with a difficult coronary artery pattern; tissue tears; leaking blood vessels; and the like.

2. *Minor events*: These occurrences do not pose safety threats individually, though – as will be seen – they have a powerful cumulative effect on compensatory ability. Examples of minor events are as follows: positioning and tensioning errors by junior surgeons; instrument-handling errors; distracting technical problems; coordination and communication problems on handover to the intensivists; insufficient monitoring during the transfer; and the like.

On average, there was one major event and six minor events per procedure. Such relatively high frequencies are commensurate with the very demanding nature of this surgical procedure. These data were derived from the 166 procedures at which there was a human factors observer present.

Fifty-two per cent of the major events were successfully compensated. Whereas over half of the major events were successfully compensated, more than 80 per cent of the minor events went uncompensated.

The Risks Associated with Events

Statistical analyses (logistic regressions) were carried out to examine the impact of major and minor events on the occurrence of death and near misses (i.e., the level 3 and level 4 outcomes). In one analysis, the contributions of the major and minor events were examined separately; the other analysis assessed their combined impact. The results are summarised below:

- The number of uncompensated major events per case was a major predictor of death ($p = 0.003$). For each uncompensated major event, the odds of death were increased by a factor of six, and the odds of death and/or a near miss were increased by a factor of 34.
- The total number of minor events per case was a significant predictor of both deaths ($p < 0.001$) and/or near misses ($p < 0.001$). Whether or not these minor events were compensated,

however, they had no significant impact upon the outcome. Minor events exerted their adverse influence in an additive fashion. Compensating for a minor event had little or no effect. What mattered was their total number per procedure.
- The occurrence of a compensated major event did not increase the risk of death. In other words, successful compensation wholly eliminated any adverse impact caused by the major event, leaving the baseline mortality odds for this procedure unchanged at around six deaths per hundred cases.

Assessing Compensatory Skills Across Surgeons

A weighted compensation score was computed for each of the 16 consultant surgeons who performed more than four arterial switch procedures during the 18-month period of the study. This score gave double weight to the percentage of compensated major events and was normalised to a 100-point scale as shown below:

$$\frac{2 \times (\text{per cent major events compensated}) + (\text{per cent minor events compensated})}{3}$$

The weighted compensation score correlated 0.96 with the per cent of major events compensated and 0.51 with the per cent of minor event compensations. Thus, the weighted value was predominantly a reflection of the proportion of major events compensated, with some recognition given to the percentage of minor event compensations. The Spearman rank order correlation coefficient between the proportion of major event compensations and the proportion of minor event compensations was 0.62 ($p < .01$, $N=16$) suggesting that the ability to compensate for events was a relatively stable characteristic across surgeons. In other words, some surgeons were consistently better compensators than others, and conversely.

Listed below are the Spearman rank order correlations with the weighted compensation score ($N = 16$):

- +0.66 ($p < .01$) with the proportion of level 1 outcomes (i.e., the best outcome).
- +0.61 ($p < .01$) with the proportion of level 1 and level 2 outcomes.

- -0.48 ($p < .05$) with the number of events (major and minor) per case.

The conclusions are clear: the best compensators get the best outcomes, but the ability to achieve successful compensation is adversely affected by the total number of events encountered in a procedure. This strongly suggests that the ability to compensate is resource-limited. Even the best compensators can only cope with so many events, be they major or minor.

Applying the Swiss Cheese Model (SCM)

We can interpret these findings with the SCM. Figure 9.2a shows the surgical defences. Figure 9.2b illustrates a typical event scenario.

We can also use the SCM to illustrate coping resources and how they get eaten away by the accumulation of minor events. Figure 9.3a shows that when – in a major event – the surgical defences are penetrated, it is still possible for the surgical team to recover the situation by a successful compensation. These recovery abilities are shown by the slice of Cheddar cheese – without

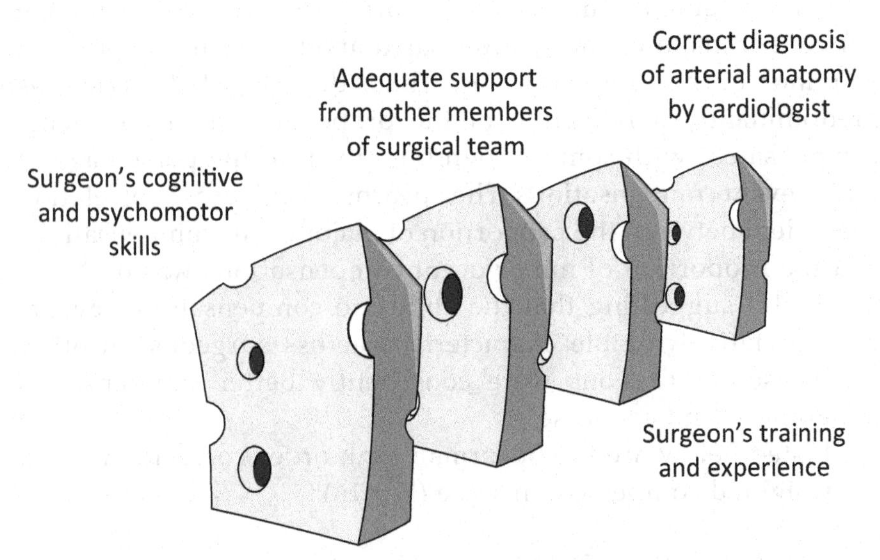

Figure 9.2a Surgical defences in the arterial switch operation represented as Swiss cheese slices in which the holes do not line up

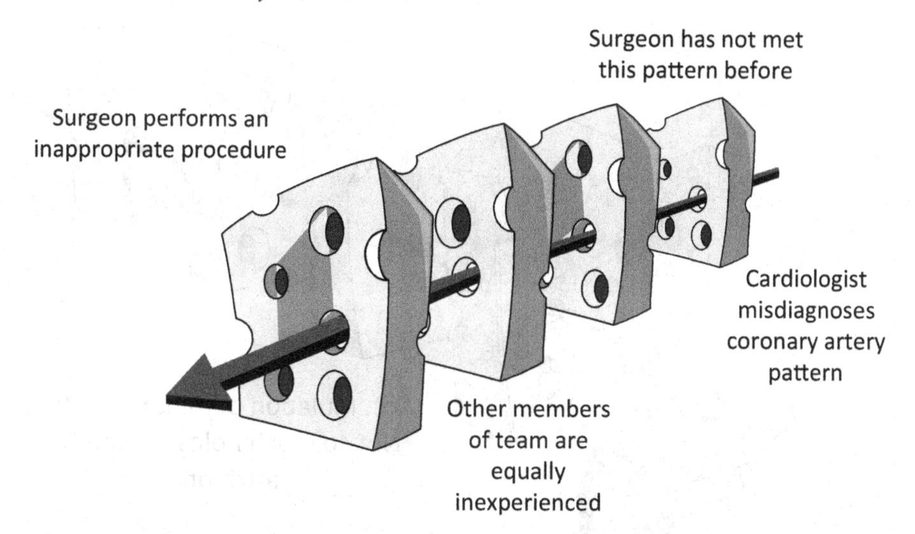

Figure 9.2b Using the Swiss cheese model to represent a typical major event scenario

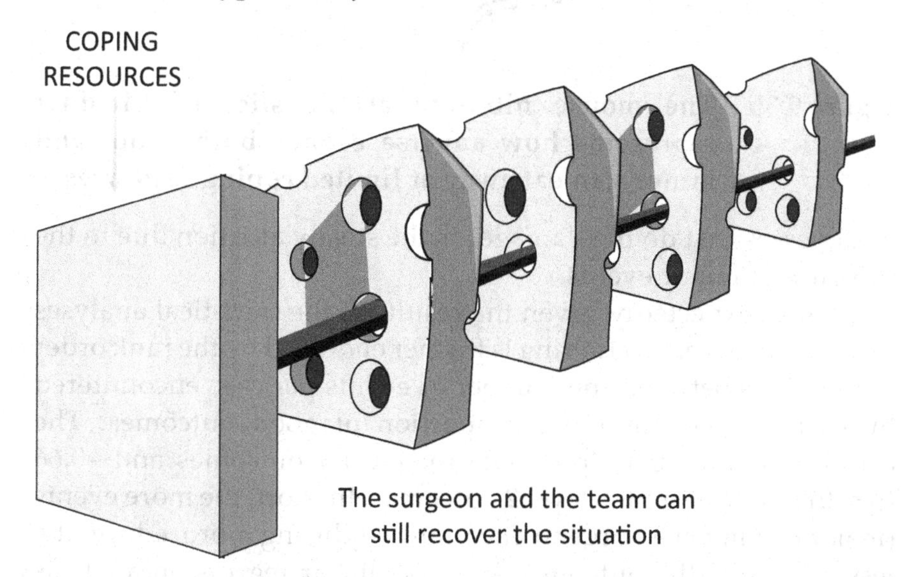

Figure 9.3a The slice of Cheddar at the left end represents coping abilities

holes – that blocks the bad event trajectory. Coping resources, unlike Emmenthale, do not have holes. Initially they are intact. But, as shown in Figure 9.3b, they can get nibbled away by the accumulated stresses associated with minor events – remember it is not whether they are compensated that is important; it is their total number during the procedure. It is not so much the holes or

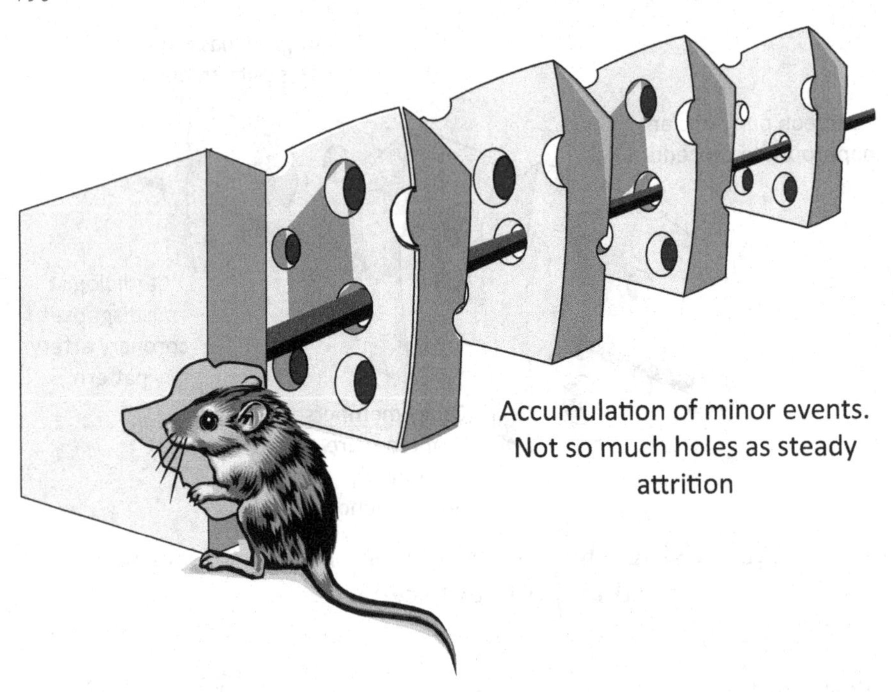

Accumulation of minor events.
Not so much holes as steady
attrition

Figure 9.3b The mouse nibbling at the slice of Cheddar represents how adverse events, both major and minor, can eat away at limited coping resources

weaknesses that do the damage, as the steady attrition due to the pressure of minor events.

Not unexpectedly, given the results of the statistical analyses discussed above, this finding is further endorsed by the rank order correlations between the number of events per case encountered by each surgeon and the proportion of good outcomes. The coefficients are –0.82 ($p < .001$) for level 1 outcomes and –0.66 ($p < .01$) for level 1 plus level 2 outcomes. In short, the more events (major or minor) a surgeon encountered during a procedure, the less good was the outcome. A simple linear regression analysis showed that events per case across surgeons (entered singly) accounted for 71 per cent of the variance in fatal outcomes ($p < 0.0004$). Clearly, the frequency of events during a procedure, irrespective of whether or not they were compensated, has a profound impact upon the surgical outcome. Taken collectively, these analyses speak strongly for the existence of finite coping resources.

Behavioural Markers of Surgical Excellence

The notion of behavioural markers of good performance was originally used to understand the characteristics of the most successful aviation crews.[11] Together, the human factors researcher and the three consultant surgeons on the research team identified the eight dimensions of surgical skill, listed below. The numbers in parentheses indicate the mean weighting score given to each factor by the three consultant surgeons on the research team (each was asked to rate the relative importance of each factor):

- *Technical skill* (26.7): the surgeon's manual dexterity, knowledge and experience.
- *Mental readiness* (14.4): how well the surgeon is prepared to carry out a particular procedure; the surgeon's belief in his ability to perform successfully; persistence in seeking a solution to a problem.
- *Cognitive flexibility* (12.8): the readiness with which the surgeon switches from one surgical strategy or hypothesis to another.
- *Anticipation* (12.8): how well the surgeon is aware of potential problems.
- *Safety awareness* (11.7): how well the surgeon handles safety-related tasks (e.g., checking prior work before moving on, controlling distraction, etc.).
- *Communication style* (8.3): how well the surgeon communicates instructions and patient-related information to other staff.
- *Team adaptation* (6.7): how well the surgeon adjusts to changes in the membership of the surgical team.
- *Situational awareness* (6.7): the extent to which the surgeon appears to have an up-to-date overview of what is going on in the operating theatre.

Identifying the Excellent Surgeons

Four out of the 16 surgeons observed scored consistently highly on all the procedural excellence indices. They made fewer errors and more compensations after adjustment for known patient risk

---

11 Helmreich, R., and Merritt, A. (1998) *Culture at Work in Aviation and Medicine: National, Cultural and Professional Influences*. Aldershot: Ashgate Publishers.

factors. Similarly, the lowest four process excellence scorers were consistent across all of the measures.

The four excellent surgeons rated highly on all the individual behavioural markers listed above. Three of them had no uncompensated major events, and the fourth surgeon was one of a group of surgeons who had a higher ratio of compensated to uncompensated events. All four surgeons were highly adaptive to unplanned departures from the normal, and they worked within teams who responded quickly to the onset of a major event.

Two surgeons, in particular, showed a considerable ability to recover minor errors made by surgical assistants. And the same two surgeons adapted their surgical and communication styles when operating with new or inexperienced surgical trainees. For example, instead of merely remarking that 'the BP's a bit iffy' as they might with an experienced colleague, they spelled out the nature of the problem in a way that made their comment unambiguous to the recipient. The positive effects of these adaptations are reflected the low numbers of uncompensated minor events for these two surgeons.

The four excellent surgeons also displayed high levels of safety awareness in making active attempts to reduce operating theatre distractions and maintain a tidy surgical field. They also showed impressive levels of cognitive flexibility and mental resilience. For example, one of these surgeons went on and off bypass six times because the patient was unstable in the immediate post-bypass period. Each time the baby was put back onto the heart–lung bypass machine, the surgeon tried out a new surgical intervention to fix one of the many underlying problems. He appeared to be working on a multiple cause hypothesis and was very flexible in his generation of hunches to explain the patient's post-bypass instability. And throughout he maintained a realistic optimism, believing on each return to bypass that the problem was soluble. Nor did he show any indications of stress.

Two of the four surgeons showed similar responses when they were surprised by unusual coronary artery patterns – that is, configurations that did not correspond to the cardiologist's pre-operative briefing. This sanguine flexibility contrasted sharply with the performance of some low-scoring surgeons who, in similar circumstances, tended to fixate on a single cause of surgical problem, and did not seem able to generate new hypotheses to resolve it. Surgeons who were not confident about their ability to solve a problem were more likely to develop cognitive 'tunnel vision' and commit 'fixation errors' when they become overly focused on a wrong surgical hypothesis.

Team and Organisational Factors

As has been shown by earlier case studies in this chapter, the performance of people at the sharp end can be influenced for good or ill by team and systemic issues. Our study revealed both positive and negative effects:

- *Positive effects* of the team and the organisation: the hospitals in which the four excellent surgeons worked always prioritised clinical goals over non-clinical ones. The cases were scheduled so that the consultant surgeon could be present in the operating theatre from start to finish of the operation. Policies were in place to reduce potential sources of distraction; for example, non-case-related communications and the number of external observers present in the theatre.
- *Negative effects:* two of the lower-scoring surgeons had a problem with team stability. In one case, there was no senior house officer specialising in paediatric cardiac surgery, and the other junior surgeon was inexperienced. There was also a different scrub nurse for each case. These difficulties meant the 'flow' of the procedure was frequently being disrupted by the need to stop and correct errors. In the other surgeon's institution, there was an ad hoc allocation of staff to theatre teams. This made it difficult to generate smooth team functioning. These latent organisational factors were particularly evident when the two surgeons encountered unexpected events. The need to correct errors disturbed the surgical flow, and this, in turn, eroded the surgeons' thinking time when faced with complication.

Concluding Remarks

There would appear to be at least two vital components underpinning heroic recoveries: a mindset that expects unpleasant surprises and the flexibility to deploy different modes of adaptation in different circumstances. In short, there is a mental element and an action element. Of these, the former is at least as important as the latter. Effective contingency planning at both the organisational and the personal levels depends on the ability to anticipate a wide variety of crises. Both components are resource-limited. Any person or organisation can only foresee and prepare for a finite number of possible circumstances and crisis scenarios. Crises consume available coping resources very rapidly. Only those people or organisations that have invested a considerable amount of preparatory effort in the pre-crisis period will be able to deploy compensatory responses in a sufficiently timely and appropriate manner so as to maintain the necessary resilience.

Chapter 10

Skill and Luck

To get into a situation from which one needs to be recovered, heroically or otherwise, usually requires a fair measure of misfortune. But there are also some recoveries that only occur through the intervention of extraordinary good fortune. Chance, as we shall see, is remarkably even-handed. Bad luck gets us into trouble, but good luck can sometimes get us out of it.

This chapter describes two aviation recoveries in which luck played a huge part. But luck was by no means the whole of it. There is a mediaeval proverb asserting that 'fortune favours the brave'. In these two cases, it also favoured flight crews with cool heads, high levels of skill, professionalism and a wealth of experience.

The Gimli Glider

Summary

On 23 July 1983, a Boeing 767 aircraft en route to Edmonton from Ottawa ran out of fuel over Red Lake, Ontario, about halfway to its destination. The reasons for this were a combination of inoperative fuel gauges, fuel-loading errors and miscalculations on the part of the flight crew – Captain Bob Pearson and First Officer Maurice Quintal. These errors and system failures are dealt with at length in the 104-page report of the Board of Inquiry[1] and will not be considered further here – except to say that among them was a diabolical sequence of errors made by a technician that defeated several defensive layers built into

1 Lockwood, Mr Justice (1985) *Final Report of the Board of Inquiry into Air Canada Boeing 767 C-GAUN Accident – Gimli, Manitoba, July 23, 1982*. Ottawa: Government of Canada.

the fuel system. Only three paragraphs are devoted to the most extraordinary feature of this event: the forced landing at Gimli, a disused military airstrip, from which all 61 passengers and eight flight crew walked away relatively unharmed (considering the alternatives) and the aircraft was fit for service after relatively minor repairs.

When the second engine stopped, the aircraft was at 35,000 feet and 65 miles from Winnipeg. All the electronic gauges in the cockpit had ceased to function, leaving only stand-by instruments operative. The First Officer, an ex-military pilot, recalled that he had flown training aircraft in and out of Gimli, some 45 miles away. When it became evident that they would not make it to Winnipeg, the captain, in consultation with air traffic control, redirected the aircraft to Gimli, now 12 miles away on the shores of Lake Winnipeg. The report continues as follows:

> Fortunately for all concerned, one of Captain Pearson's skills is gliding. He proved his skill as a glider pilot by using gliding techniques to fly the large aircraft to a safe landing. Without power, the aircraft had no flaps or slats to control the rate and speed of descent. There was only one chance of landing. By the time the aircraft reached the beginning of the runway, it had to be flying low enough and slowly enough to land within the length of the 7,200 foot runway.
>
> As they approached Gimli, Captain Pearson and First Officer Quintal discussed the possibility of executing a side-slip to lose height and speed close to the beginning of the runway. This the Captain did on the final approach and touched down within 800 feet of the threshold.[2]

The last sentence of this laconic account is a masterpiece of understatement. Side-slipping is a manoeuvre that puts an aircraft into an off-balance mode of descent with one wing raised and advanced. If carried out properly, it allows the plane to descend rapidly without gaining excess speed. It is the kind of manoeuvre that skilled pilots use to put gliders and light aircraft into small fields in a hurry. It is unlikely that either Boeing or Captain Pearson's employers had ever imagined it being applied to a wide-bodied jet airliner. As it turned out, however, it was almost certainly the only way that the aircraft could have made a safe landing under those circumstances.

---

2 Ibid, p. 29.

Running Out of Fuel

The flight crew and the passengers had just finished dinner when the first warning light came on indicating a fuel pressure problem on the left side. Air Canada Flight 143 was heading westwards over Red Lake Ontario at 41,000 feet and 469 knots at the time. At first, they thought it was a pump failure problem, but seconds later warning lights indicated that there was a loss of pressure in the right main fuel tank. The captain quickly ordered a diversion to Winnipeg Airport, 120 miles away. It was clear that they were running out of fuel.[3]

The left engine was the first to flame out. Twenty minutes after they became aware of a problem and when their altitude was 28,500 feet, the right engine stopped. They were 65 miles from Winnipeg. The auditory warning system gave out a loud 'bong'. This was the 'all engines out' sound. The pilots had never heard it before. It was not a part of the simulator training. Starved of fuel, both Pratt and Whitney engines had flamed out. The 156-tonne, state-of-the-art Boeing 767 had become a glider. The auxiliary power unit (APU) designed to supply electrical and pneumatic power under emergency conditions and fed from the same fuel tanks as the main engines had also packed up.

Approaching 28,000 feet, the 767's glass cockpit went dark. Most of the instrument panels went blank. Captain Pearson was left with a radio and standby instruments, but no vertical speed indicator – the glider pilot's instrument of choice. A magnetic compass, airspeed indicator and an altimeter were the only instruments still functioning. Hydraulic pressure was falling fast and the aircraft's controls were rapidly becoming inoperative. But the Boeing engineers had foreseen this most unlikely of scenarios and had provided one last defence: the Ram Air Turbine (RAT).

The RAT dropped from near the right wheel housing and used wind power to turn a four-foot propeller, providing just enough hydraulic power to control the ailerons, elevators and rudder and enable a dead-stick landing.

---

3 Principal sources: Nelson, W. (1997) *The Gimli Glider*. (http://www.wadenelson.com/gimli.html). Williams, M. (2003) 'The 156-tonne Gimli Glider'. *Flight Safety Australia*, July–August, 22–27. Wikipedia (2008) *Gimli Glider* (http://en.wikipedia.org/wiki/Gimli_Glider).

As Captain Pearson began gliding the large wide-bodied jet, Maurice Quintal started scanning the manuals for procedures dealing with the loss of both engines. There were none. Neither of the two pilots, nor any other 767 pilot, had ever trained for this most unlikely event.

Lucky break number one: Captain Pearson was an experienced glider pilot and part-owner of a sailplane that he flew at weekends. This made him familiar with some flying techniques that were almost never used by commercial pilots. In order to achieve the largest choice of possible landing sites, he needed to fly the 767 at the 'best glide ratio speed'. His educated guess was 220 knots. The windmilling fan blades on the stopped engines created enormous drag, giving the aircraft a sink rate of between 2000 and 2500 feet per minute. First Officer Quintal started making glide-slope calculations, and both he and the Winnipeg controllers decided that they were dropping too fast to make Winnipeg. Pearson trusted Quintal's judgement completely at this critical moment and turned the aircraft southwards.

Lucky break number two: Quintal selected Gimli as their landing site. This was the site of an abandoned Royal Canadian Air Force Base, and Quintal had been stationed there during his national service. This was very fortunate indeed, because the airfield was not listed in Air Canada's airport manuals. As far as anyone knew, Gimli's two 6,800-foot runways (32L and 32R) would be deserted. But that was far from the case.

Runway 32L had been partially converted into a two-lane dragstrip. A steel guard rail had been installed between the two strips. This was the runway that Captain Pearson would attempt to land on. On this particular July afternoon, drag racing was about the only car racing event that was not taking place on the runway. It was 'Family Day' for the Winnipeg Sports Car Club, and go-cart races were being held on one portion of 32L. Around the edges of the runway were large numbers of cars, campers, children and families. It had all the makings of a disaster.

The Approach and Landing

Approaching Gimli, the two pilots made an unpleasant discovery. The RAT did not supply hydraulic pressure to the 767's landing gear. Pearson ordered a 'gravity drop'. Quintal hit the button to release the gear door pins, and they heard the main gear fall and lock into place. But the instruments showed only two green lights, not three. The nose gear had gone down but had not locked. As it turned out, the absence of a nose wheel saved lives.

Six miles out, Pearson began his final approach on 32L. As he came nearer to the threshold, the captain realised he was coming in too high and too fast and would overrun the runway at its present speed, and there was no way of applying reverse thrust. So Captain Pearson took a gamble that the huge 767 would respond in the same way as a smaller aircraft and executed a sideslip by turning the yoke to the right at the same time as he jammed his foot against the left rudder pedal. The aircraft was manoeuvred into a steep angle and rapidly lost altitude. Some of the passengers looked up at nothing but blue sky, and others stared at the golf course below. Quintal, in the right hand seat, looked down, rather than sideways at the captain.

Captain Pearson did not kick the aircraft straight until the very last moment, at an indicated altitude of 40 feet. Why did he not land on 32R? Captain Pearson said later that he never even saw 32R, focusing instead upon airspeed, attitude and the plane's relationship to the threshold of 32L.

The sideslip manoeuvre required exceptional piloting skills since the indicated airspeed was incorrect because the angle of the aircraft was not the same as its direction of travel. It came down to the captain's judgement and experience as a glider pilot.

As the 767 silently levelled off and the main gear touched down, spectators, racers and kids on bicycles fled the runway. Pearson stood on the brakes the moment the main gear touched down. Explosions shook the cockpit as two tyres blew. The unsecured nose wheel collapsed with a bang, slammed against the tarmac, and then began throwing a 300-foot shower of sparks. The right engine nacelle struck the ground. And as a final touch of finesse, Pearson applied extra right brake so that the main gear would straddle the central guard rail. The 767 finally came to a stop less than 100 feet from spectators, barbecues and campers. All of the racers had managed to escape the path of the aircraft.

The fuselage was intact. For an instant the passengers were silent, and then cheers and applause broke out. The only injuries occurred when passengers, exiting a rear door down a too-steep escape chute, hit the tarmac. No injuries were life-threatening.

The time of the landing was 20.38 hours. Just 17 minutes had elapsed since Captain Pearson had glided the 767 from 28,500 feet to a safe landing.

The aircraft (then designated as Air Canada Aircraft #604) was repaired sufficiently to be flown out of Gimli two days later. After approximately one million dollars in repairs, the 767 re-entered the Air Canada fleet and flew its last revenue flight on 1 January 2008. Captain Pearson retired in 1993 from Air Canada and raised horses. Maurice Quintal went on to captain 767s, including the 'Gimli Glider'. In 1985, the two pilots were awarded the first ever Fédération Aéronautique Internationale Diploma for Outstanding Airmanship.

Captain Al Haynes and United 232

Summary

On 19 July 1989, a McDonnell Douglas DC-10, United Airlines Flight Number 232, was en route from Denver to Chicago. The aircraft was commanded by Captain Al Haynes, a pilot with 30,000 hours of flying experience. On the flight deck with him were First Officer William Records and Flight Engineer Dudley Dvorak. There were also eight flight attendants. After the explosion in the number two engine, the pilots were greatly assisted by Captain Dennis Fitch, a United Airlines training and check pilot with over 3,000 hours on the DC-10. He had been flying as a passenger on United 232.

At 15.16 hours, the aircraft suffered an almost unimaginable misfortune.[4] While flying at cruise altitude, the tail-mounted number two engine's fan rotor disintegrated and cut through all three of the aircraft's hydraulic systems. The probability of losing all three hydraulic systems was considered by the designers to be less than one in a billion (10^{-9}) and no emergency procedures

4 Haynes, A.C. (1992) 'United 232: coping with the loss of all flight controls.' *Flight Deck*, 3: 5–21.

were available to cover this almost unthinkable possibility. Al Haynes described their plight as follows:

> That left us at 37.000 feet with no ailerons to control roll, no rudders to co-ordinate a turn, no elevators to control pitch, no leading edge devices to help us slow down, no trailing edge flaps to be used in landing, no spoilers on the wings to slow us down in flight or help braking on the ground, no nosewheel steering and no brakes. That did not leave us a great deal to work with.[5]

Forty-five minutes later, the aircraft crash-landed at Sioux City, Iowa. Of the 285 passengers and 11 crew members on board, 174 passengers and ten crew members survived. About 14 seconds into the emergency, the aircraft had reached 38 degrees of bank to the right, on its way to rolling over on its back and then falling out of the sky. At this point, Al Haynes took control of the aircraft and slammed the number one (left engine) throttle closed and opened the number three (right engine) throttle all the way, and the right wing started to come back up. After that, the two pilots – plus Dennis Fitch, flying as passenger – learned to use the differential thrusts of the two remaining wing-mounted engines to skid the aircraft left or right in descending turns. In this way, they coaxed it into Sioux City and might have landed safely had the aircraft not been affected by uncontrollable manoeuvres (phugoids) while low on the approach.

To save the lives of 184 out of a total of 296 people aboard – 296 people who were almost certainly doomed to die in an uncontrollable crash – was a remarkable feat of airmanship. But there were other factors as well. What follows is largely taken from Captain Haynes's own account of the factors that contributed to the remarkable survival of more than half of the people on board.

Luck, the Number One Factor

Captain Haynes attributed the partial survival of United 232's passengers and crew to 'an unbelievable amount of good luck'.[6] Some the factors that shaped his opinion are listed below:

- The aircraft was over the relatively flat lands of Iowa rather than the Rockies or the Pacific Ocean. While Al Haynes had doubts

5 Ibid, p. 7.
6 Ibid, p. 7.

about reaching the airport at Sioux City, he had the feeling that if they could just get the aircraft on the ground, there would be survivors because of the flat farmland below. This gave the flight crew the idea that the situation was not wholly impossible.

- There were no seasonal thunderstorms in the location. As Captain Haynes wrote later: 'If you have ever flown over the US Midwest in July, you know there is usually a line of thunderstorms that runs for the Canadian border down to Texas, and it would have been absolutely impossible with marginal control of the aircraft to get through a thunderstorm safely.'[7]
- The timing of the accident coincided with shift changes in the local hospitals, so that both the morning and the evening shifts were available. The same was true of all the other emergency services around the Sioux City area. There were so many volunteers from the various emergency units and health clinics in the vicinity that the hospitals had to turn some of them away.
- It was the only day of the month in which the 185th Iowa Air National Guard was on duty, and that meant 285 additional trained personnel waiting for the arrival of the aircraft at Sioux City.
- But perhaps the luckiest factor of all was the quality of the people whose task it was to deal with the emergency: the flight crew, the flight attendants, the air traffic controllers, the emergency services and the United Airlines ground staff. The odds against getting such an outstanding group of individuals in one place at one time to deal with this singular emergency must have been immensely large.

Communications

Captain Al Haynes placed good communications second on his list of the factors that contributed to the survival rate on United 232. These started in the cockpit with communications to air traffic control in Minneapolis Centre and ended with the Sioux City approach control and control tower. Here is a slightly abridged version of the last interchanges between United 232 and Kevin Bauchman, the air traffic controller who became their primary

7 Haynes, A. *Eyewitness Report: United Flight 232* (http://www.airdisaster.com/eyewitness/ua232.shtml).

contact with the Sioux City tower.[8] They cover the last twelve minutes of the flight.

On making contact with Sioux City, Al Haynes explained that they had minimal flight controls and could only make right turns using power. After establishing the controllability of the aircraft, how many people there were on board and the fuel remaining, the controller asked the flight crew to turn on to a heading of 240 degrees:

[15:47]

Sioux City: United 232 heavy, you're going to have to widen out just slightly to your left, sir, to make the turn to final, and also to take you away from the city.

UAL 232: Whatever you do, keep us away from the city.

Sioux City: 232 heavy, be advised there is a four-lane highway up in that area, sir, if you can pick that up.

UAL 232: Okay, we'll see what we can do here. We've already put the gear down, and we're going to have to be putting it on something solid if we can.

Sioux City: United 232 heavy, roger, airport's currently at you one o'clock position, 10 miles.

Sioux City: If you can't make the airport, sir, there's an interstate that runs north to south to the east side of the airport. It's a four-lane highway.

UAL 232: We're just passing it right now. We're going to try for the airport.

Sioux City: United 232 heavy, roger, and advise me when you get the airport in sight.

UAL 232: We have the runway in sight, and will be with you shortly. Thanks a lot for your help.

Sioux City: United 232 heavy, winds currently 360 at 11. Three sixty at eleven. You're clear to land on any runway.

UAL 232: You don't want to be particular and make a runway, huh?[9]

Sioux City: . . . 010 at 11, and there's a runway that's closed, sir, that could probably work too. It runs northeast to southwest.

UAL 232: We're pretty much lined up on this one here.

---

8 Recording provided by NASA-Dryden, 1991.

9 Even under these conditions, Al Haynes can make a joke. He made several others during the emergency.

Sioux City: United 232 heavy, roger sir. That's a closed runway, that'll work, sir. We're getting the equipment off the runway and they'll line up to that one.

UAL 232: How long is it?

[16:00]

Sioux City: At the end of the runway is just a wide open field, sir, so the length won't be a problem.

UAL 232: Okay. [final communication]

Al Haynes later commented on these interchanges:

> If you have a serious problem like we did, and you need the kind of help that does not add to the tension level, a voice like Bauchman's, as calm and steady as he was, certainly was an influence on us and helped us remain composed... When I had the opportunity later to compliment Bauchman on his coolness throughout the tense situation, he told me, to my surprise, that he had transferred to Sioux City because he found his previous duty station too stressful.[10]

Preparation

This was the third item on Al Hayne's list of factors that contributed to their salvation. The benefits of preparation took a number of forms:

- The Emergency Response Group in Sioux City had a disaster drill they practised every three years. During the 1987 exercise, the organisers came up with a prophetic scenario: a wide-bodied jet that did not serve Sioux City crashed on the airport's closed runway. The Director of Emergency services was not happy with the exercise and revised the plan in order to bring more services and to involve more of the small communities in the area. And that's what happened on the day that United 232 came down. The organisers rehearsed it and drilled at all there to ensure that they were properly prepared for any reality. They were.
- United Airlines put their cabin crew through a recurrent training programme every year in which they are taught how to prepare passengers for an emergency landing. The United 232 cabin crew had been practising these drills for the whole time they had been employed, from one month for the most junior flight attendant to between 15 and 20 years for the most senior ones.

10 Ibid, p. 4.

- Following the Tenerife disaster, due in large part to the social dynamics on the flight deck of the KLM 747, United Airlines was one of the first to institute Crew Resource Management (CRM) aimed at breaking down the hierarchy that often existed between a senior captain and the First and Second Officers. The programme began in 1980, and was initially called Command Leadership Resource Management (CLRM) training. Its main purpose was to teach flight crews to work together so that all members of the flight crew felt free to express and opinion when problems arose. Al Haynes made the following comments after the event:

> 'The bottom line for pilots is that you have resources available to you, use them as team members – you are not alone up there. If you do have a co-pilot, listen to her or to him. They are sure to have some advice for you. There were 103 years of flying experience in that cockpit when we faced our end and they cam through to help – but not one minute of those 103 years had been spent operating and aircraft the way we were trying to fly it. If we had not worked together. With everybody coming up with ideas and discussing what we should do next and how we were going to do it, I do not think we would have made it to Sioux City.'[11]

Execution Came Next

During the course of the emergency, the flight crew had to cope with a succession of novel problems. Some of these are discussed below:

- First Officer Records was flying the aircraft on autopilot when the number two engine disintegrated. Without any warning there as a very loud explosion. The pilots' first thought was that it was a decompression, but there was no rush of air and no change of pressure. Bill Records cut the autopilot off and switched off the autopilot. Fourteen seconds into the incident he said 'Al, I can't control the airplane.' Al Haynes noticed that Records had applied full left aileron and had the yoke completely back in his lap, calling for full up-elevator. But the aircraft was in a descending right turn and at an increasing angle.
- Meanwhile, Second Officer Dudley Dvorak was reading out the check list for shutting down the number two engine. The

11 Ibid, p. 6.

first item was 'close the throttle'. But the throttle would not close. The second item was to close off the fuel supply to the engine, but the fuel lever would not move. Then Dudley said to actuate the firewall shut-off valve. When Al Haynes did that, the fuel supply was finally shut off.

- As the aircraft reached about 38 degrees of bank on its way to rolling inverted, Al Haynes slammed the number one (left) throttle closed and firewalled the number three throttle, and the right wing came slowly back. Al Haynes was later asked how they thought to use the differential thrust of the two surviving wing-mounted engines to control the aircraft. He answered: 'I do not have the foggiest idea. There was nothing left to do, I guess, but it worked. Here is another instance where I talk about luck; we tried something that we did not know what to expect from and we discovered that it worked'.[12]
- After 15 minutes, the flight crew were advised that Captain Dennis Fitch was a passenger in first class, and they invited him on to the flight deck. For the next 30 minutes, Fitch operated the one and three throttle levers keeping the aircraft on a moderately even keel while they approached Sioux City. He turned out to be remarkably good at it.

The Final Step Was Cooperation

Captain Haynes is lavish in his praise about the cooperation that existed within and between several teams: the members of the flight crew and Denny Fitch; the high level of cooperation between the cabin and the flight crew; between the flight crew and air traffic control; and between the emergency services, the hospitals and the National Guard. He was also very impressed with the way that United Airlines responded to the Sioux City crash. They pulled large numbers of staff from their jobs in San Francisco and Seattle and sent them to Sioux City. By the middle of the next day, there was at least one United Airlines employee for every family there.

---

12 Ibid, p. 6.

Al Haynes on Post-traumatic Stress Disorder

Let me give Al Haynes the last word on the United 232 event:

> This is my 52nd speech on 232. Every time I give it, I think I convince myself just a little bit more that there was nothing else I could do. And it's part of my healing process. To not talk about it, to bury your head in the sand and pretend it didn't happen, you're going to explode someday. So if somebody wants to talk about a trauma, listen to them. If they want you to talk, you talk about it, and listen to them. But be there for them, and help them. That's very important.'[13]

A specialist on post-traumatic stress disorder could hardly have expressed it better.

Final Word

Given the remarkable qualities of the pilots discussed in this chapter, you may well think that I have overestimated the part played by good luck. Happy chance of one kind or another plays a role in all heroic recoveries. But consider the following: what are the odds that the same flight deck contains, when both engines cut out, a captain with the gliding skills of Bob Pearson and a co-pilot like Maurice Quintal who has actually been stationed in a decommissioned and now unrecorded military airfield that is conveniently close by?

What are the chances that all of the things mentioned by Al Haynes come together at the right time – flat farmlands below, no seasonal thunderstorms, both hospital and emergency shifts available at the crash site, the presence of the Iowa National Air Guard, the exercises carried out by the 'Siouxland' emergency services, Denny Fitch as a dead-heading passenger and an air traffic controller of the quality of Kevin Bauchman to see you down? The answer must be, on both counts, vanishingly small. This in no way diminishes from the extraordinary professionalism of the people concerned, but luck was a major factor in both cases. May we all be so lucky.

---

13 Ibid, p. 19.

Chapter 11
Inspired Improvisations

Although many of the recoveries discussed so far have involved improvisations that were truly inspired – Bob Pearson's side-slipping and Al Haynes's use of the throttles, to name but two – they were just part of a number of other contributing factors. But in the two cases discussed here the inspired improvisations were the defining features, the ones by which the episodes are best remembered. This is not to say that skill, experience, professionalism and the like played no part, of course they did; but they stood as a backdrop to the brilliant and novel adaptations that were conjured up by the main players.

General Gallieni and the Paris taxis

Background

In the closing days of August, 1914, Paris was not a good place to be. It stood directly in the path of General Alexander von Kluck's largely victorious First Army, the largest and most westerly of the six German armies that had invaded Belgium and north-eastern France at the beginning of August. Since then, von Kluck had driven the French 5th and 6th Armies, and their allies, the British Expeditionary Force (BEF), relentlessly southwards. On 30 August, von Kluck's forces were less than 40 miles to the north of Paris. General Joffre, commander-in-chief of the French armies, had effectively given up the idea of defending Paris and the French government was about to flee to Bordeaux.

It was at this point that von Kluck made a fateful and ultimately flawed decision. The German plan of attack (the Schlieffen Plan)[1]

1 The brain child of Count von Schlieffen, Chief of the General Staff, originally prepared in the late 19th century and subsequently modified by von

First Army to pass south of Paris and then to turn east towards the Vosges. But von Kluck chose to swing eastwards while he was still north of the capital. He believed that neither the French nor the British were in a fit state to offer any serious opposition. But he had not counted on the ingenuity of General Gallieni, the military governor of Paris.

General Gallieni

In the summer of 1914, Joseph-Simon Gallieni was a 65-year-old professional soldier, educated at the Saint-Cyr military academy. The greater part of his career was in the French colonies, first in West Africa, then Martinique and Upper Senegal. From 1896 until 1905, he was governor-general of the newly acquired island of Madagascar. On his return to France he was made a general of a division and was appointed military governor of Lyon. At the beginning of the First World War, he was chosen to head the military government of Paris.

Although von Kluck's inward wheel was initiated on the morning of 31 August, the significance of the manoeuvre was not immediately grasped by Joffre and the French High Command. But they did agree to bring a corps each from the 1st and 2nd Armies, fighting on the Moselle line in the east, to boost the Paris garrison.

Gallieni received definite news of the turn from an aviator on the morning of 3 September. He immediately saw the possibility of attacking von Kluck's exposed flank. But Gallieni had to persuade Joffre[2] to initiate the attack, and Joffre in turn had to cajole the reluctant British to support them. This took time and caused Gallieni a great deal of frustration.

Moltke, the German supreme commander. It was a plan to defeat France in the six weeks it would take the Russians to mobilise. In the event, the Russians mobilised in a shorter time, but they were decisively beaten at Tannenberg in East Prussia on 31 August. The Schlieffen Plan ceased to function when von Kluck's south-easterly turn precipitated the Battle of the Marne, after which the Germans retreated.

2 This is not to diminish General Joffre's huge contribution to the outcome of the First World War. He was a 'three-square-meals-a-day-and-in-bed-by-ten-o'clock-man' and it was his sublime imperturbability that played a very large part in withstanding the onslaught of the German invasion.

Some More Background: the Battle of the Marne

This French victory – the British did not really shine here – put an end to the Schlieffen Plan and the war of rapid movement that had prevailed since the early days of August. In its wake, trenches were dug from the Channel coast to the Swiss border, and the fighting settled down to four bloody years of a largely unwinnable contest that only ended when the Americans arrived and the German civilian infrastructure collapsed in 1918.[3]

The battle started on 5 September when General Manoury's 6th Army (a core part of the Army of Paris that Gallieni now commanded) accidentally stumbled onto the leading elements of von Kluck's First Army. Under pressure from his commander, von Moltke, General von Kluck had earlier abandoned his south-easterly advance, and had swung round to face the French 6th Army threatening his exposed flank. In doing this, he had opened a 30-mile wide gap between his First Army and the German Second Army. The French and the BEF sought to exploit this opportunity, but the British, in particular, advanced very slowly. Meanwhile, Manoury's 6th Army, was under furious attack and was beginning to fall back. The moment of crisis had arrived; and so had Gallieni's lasting place in military history.

Before leaving this overview of the Battle of the Marne, let me quote General von Kluck's generous comment on his main opponent, the French. Later, he wrote about 'the reason that transcends all others' as to why the Germans failed at the Marne.

> [It is] the extraordinary and peculiar aptitude of the French soldier to recover quickly. That men will let themselves be killed where they stand, that is a well known thing and counted on in every plan of battle. But that men who have retreated for ten days, sleeping on the ground and half dead with fatigue, should be able to take up their rifles and attack when the bugles sound, is the thing upon which we never counted. It was a possibility not studied in our war academy.[4]

3 I have often thought that had the Marne engagements turned out differently in September 1914, the twentieth century might have taken quite another and perhaps kinder course – no brutal stalemate on the Western Front, no Treaty of Versailles, and perhaps no breeding ground for the rise of National Socialism in post-war Germany. But I'm getting beyond my remit – and my competence.

4 Suchman, B.W. (1962) *The Guns of August*. London: Four Square Edition, (p. 485).

The Marne Taxis

Between 6 and 8 September, Manoury's 6th Army was hard-pressed by von Kluck who was close to a breakthrough. Earlier Manoury had asked Gallieni what his line of retreat would be should his army be overwhelmed. 'Nowhere,' Gallieni answered.

Help was at hand. On 6 September, units of the 7th Division, sent from the eastern frontier, were arriving at railway sidings in the Paris area. But there were no army motor vehicles or drivers to take them to the fighting.

It is easy to understand Gallieni's exasperation at this moment. He, rather than Joffre, had seen the military opportunity of attacking von Kluck's exposed flank. It was he who had spent the past few days trying to persuade the French High Command and the reluctant British to turn around[5] and take the offensive. But all he had managed to wrest from Joffre in tangible terms was the transfer of the 7th Division from the eastern frontier to Paris. Now they had arrived; but there was no military transport to extricate them from the choked French rail system and move them to the beleaguered front line on the Ourcq. Then Gallieni had his brain wave: 'Why not use taxis?'

Perhaps it was his many years as a colonial administrator that allowed him to go beyond the conventional military mindset and consider using civilian transport (the numbers of the taxis were later swelled by buses, limousines and stray trucks). Whatever the reasons, it was a brilliant piece of lateral thinking, the true basis for the 'miracle on the Marne' – I mention this because some of the more spiritually minded French believed that it was Joan of Arc's influence at work here.

Several Paris taxis were already in the service of the military government of Paris. General Clergerie, Gallieni's chief of staff, estimated that with 500 more, each carrying five soldiers and making the 60-kilometre round trip twice; he could transport 6,000 troops to the embattled 6th Army on the River Ourcq. The order was issued at 1 pm, and the time of departure fixed for 6

5 This was something of an understatement. Field Marshal Sir John French, commander of the BEF, had already resolved to retreat to the Channel ports and go home. Fortunately, General Kitchener, his boss, told him to stay put and obey Gallieni's orders. Sir John French was not the greatest commander that the British had ever fielded.

pm that evening. The police passed the word to the taxi drivers who emptied out their passengers, explaining proudly that they had to go to war. After returning to their garages to fill up with petrol, they began assembling at the Esplanade des Invalides, close to Gallieni's headquarters. By 6 pm all 600 (mainly Renault AGs) were lined up in perfect order.

Called to inspect them, Gallieni was delighted and exclaimed. '*Eh bien, voila a moins qui n'est pas banal!*' (Well, here at least is something out of the ordinary!) 'What about the fare?' asked one of the taxi drivers, a true Parisian. Compensation eventually did materialise at 27 per cent of the meter reading.

By 8 September, two days after the taxis were first employed, several thousand fresh troops had reinforced Manoury's hard-pressed 6th Army. Under pressure from both the French 5th and 6th Armies, the German Armies withdrew to the Aisne. The Battle of the Marne was over – a French victory after ten days of defeat and retreat. Without the taxis and the 6,000 soldiers they brought from Paris, it is unlikely that Manoury's 6th Army could have held its ground. Paris was never threatened again throughout the First World War. Gallieni had saved Paris.

In 1915, Gallieni was made minister of war in the cabinet of Prime Minister Aristide Briand, but ill health caused him to resign shortly before his death in 1916. The title of marshal was awarded posthumously to Gallieni in 1921.

Captain Gordon Vette and the Rescue of Jay Prochnow

I have had the privilege of meeting and communicating with Captain Vette on a number of occasions. In 2003, I was also very honoured to be a recipient of The Captain A.G. Vette Flight Safety Research Trust Award. The citation read 'Awarded to Professor James Reason in recognition of an outstanding contribution to flight safety enhancement within the aviation industry and for his work and research in human factors and the design of more error tolerant systems.' Please allow me this small boast; it's not often that an academic psychologist gets recognised by an internationally acclaimed aviator – Gordon was a fan of the Swiss cheese model. I hope to explain something of the grounds for his acclaim below.

Background

Three days before Christmas, 1978, Jay Prochnow, a former US Navy pilot, took off from Pago Pago, American Samoa, for Norfolk Island, lying several hundred miles north of New Zealand. The distance was 1475 nautical miles and the journey was expected to take 14 hours, the aircraft having a fuel endurance of around 22 hours. He was ferrying a Cessna 188 Agwagon (a crop duster) from California to Australia.

Toward the latter portion of his flight, Prochnow received the low-frequency beacon at Norfolk on his automatic direction finder (ADF)[6] and was encouraged to see that the ADF needle pointed straight ahead, indicating that he was on the home stretch. But Norfolk Island failed to appear at the expected time. Assuming that the wind was more adverse than anticipated, he pressed on. However, the ADF needle began to swing and drift. He tried to tune it into other beacons, but these provided similar irrational readings. It was clear that the ADF had failed. Prochnow realised he was lost, a bad thing to be when you're flying over the Pacific Ocean in a crop duster without a radio directional finder. After using his Navy training to conduct a square search,[7] he was still lost and very worried. He sent a 'Mayday' call to air traffic control at Auckland using his high-frequency (HF) radio, declaring a serious emergency.

On the same day, an Air New Zealand DC-10, commanded by Gordon Vette with First Officer Arthur Dowey as his co-pilot, had taken off from Nandi in Fiji and was en route to Auckland. Auckland air traffic control estimated that his aircraft was the closest to the Cessna, and, after consulting his passengers, he agreed to divert with the hope of finding Prochnow. Both pilots had navigation licences and since the aircraft was tankering fuel, Gordon Vette had sufficient reserves to engage in the search – and three inertial navigation systems as well.[8]

---

6 An ADF will always point directly to the broadcast transmission tower, irrespective of the ADF's attitude or heading.

7 A square search involves flying a pattern of ever-widening squares with the hope of finding a target before the fuel is exhausted.

8 The main sources for the Prochnow rescue were: 'Mayday in December' (http://www.navworld.com/navcerebrations/mayday.htm); BALPA's Outstanding Airmanship Awards (Harlington: BALPA); and Schiff, B. 'Ferry

Finding Jay Prochnow

It was now late in the day and the sun was beginning to set. Prochnow was thinking of ditching his plane into the ocean while there was still enough light left. Had he done that, it is almost certain that he would not have survived. Prochnow had now been flying for nearly 20 hours and was long overdue at Norfolk Island.

At around this time, Gordon Vette made radio contact with the Cessna and began a series of remarkable navigational exercises in order to establish the relative positions of the two aircraft. The first thing he asked Prochnow to do was to turn his aircraft towards the setting sun and to report his heading. Prochnow reported his heading as 274 degrees magnetic. Gordon did the same thing in his DC-10 and established a magnetic heading of 270 degrees. This put the Cessna south of the DC-10.

Then Gordon asked Jay Prochnow to place his hand on the instrument panel and to count the number of fingers between the horizon and the centre of the sun. Prochnow established the elevation of the sun as four fingers. Gordon Vette repeated the exercise on his flight deck and came up with elevation of two fingers. Since the elevation of the sun, as measured by Prochnow was higher than that established by Gordon Vette, the Cessna was closer to the sun, or west of the DC-10. Given that each finger subtended an angle of just over two degrees, with each degree worth 60 nautical miles, Gordon estimated that the Cessna was around 240 nautical miles to the south-west of the DC-10.

Soon after this, the DC-10 established VHF contact with the Cessna. Prochnow was asked to fly due east towards the DC-10. Then Gordon had another brain wave; he recognised that the VHF link could be used to establish the position of the Cessna. He asked Prochnow to orbit and to keep talking. He knew the diameter of the VHF range circle was 400 nautical miles. Flying his big jet across the circle, he could note the points when he made and lost contact (points 1 and 2 on Figure 11.1). After losing contact (at point 2), he turned 90 degrees to the left and started his aural box pattern (see Figure 11.1). After flying on this new

---

Pilot, aka Help from Above' (http://www.flight.org/forums/showthread.php?t=269110).

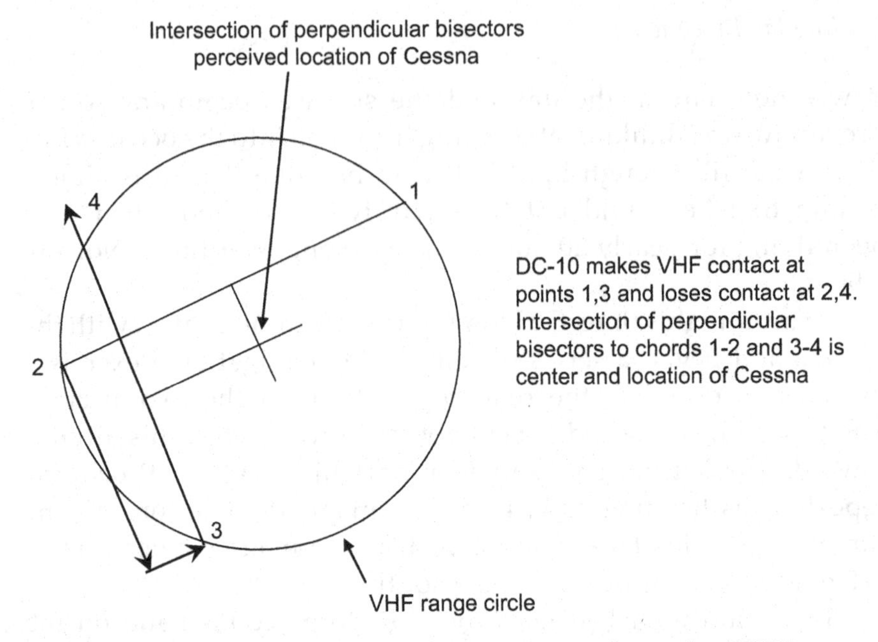
Intersection of perpendicular bisectors
perceived location of Cessna
1
2
3
4
DC-10 makes VHF contact at
points 1,3 and loses contact at 2,4.
Intersection of perpendicular
bisectors to chords 1-2 and 3-4 is
center and location of Cessna
VHF range circle

Figure 11.1 Flying the aural box pattern into the VHF range circle

leg for a reasonable time, he turned left and then immediately left again (point 3), each turn being through 90 degrees, and continued until he lost contact once again (point 4). He then drew two lines perpendicular to the chords he had flown within VHF range circle. Where they intersected was the centre of the range circle, and that's where the Cessna should be. But it wasn't there – probably due to small timing errors in the aural box procedure.

However, Gordon Vette had one more trick up his sleeve. The sun was just disappearing over the horizon. It was possible to establish the difference in longitude by noting the times at which the sun disappeared at both Norfolk Island and aboard the Cessna. The Cessna's time was adjusted for altitude as Prochnow would see the sun disappear later than at sea level. The difference between the times was 22.5 minutes which equates to 5.6 degrees of longitude (a degree is equal to 4 minutes in time). Knowing the coordinates of the observers on Norfolk Island, the DC-10 crew calculated that the Cessna was 291 nautical miles east of Norfolk Island.

Prochnow was directed to fly north-west. The Cessna had now been airborne for 20.5 hours and had very little fuel left. Then Prochnow saw an oil rig under tow. They radioed their coordinates to the DC-10 crew who were able to rendezvous with the Cessna – the Cessna was spotted by a passenger. They were now less than 150 nautical miles from Norfolk Island. Gordon gave Prochnow a magnetic heading of 294 degrees. Jay Prochnow landed safely close to midnight having been flying for more than 23 hours – he had stretched his 22 hours of fuel by skilful use of the cruise control.

Afterwards

The extraordinarily lucky Jay Prochnow is now (I believe) an airline captain. I suspect that very few DC-10 pilots on a rescue mission could have pulled all those life-saving rabbits out of the hat. Gordon Vette was a navigational magician: a supreme airman. His actions were the epitome of inspirational improvisations: one problem after another, one ingenious solution after another.

But there was a further chapter in Gordon Vette's career with Air New Zealand – after 33 years of flying at the time of the Prochnow rescue, he was only a few years short of retirement. But 1979 was a very sad year for New Zealand. It was in late November that the Air New Zealand DC-10 crashed into Mt Erebus on a sightseeing flight to Antarctica (discussed in Chapter 7). There were no survivors.

An Erebus Postscript

Why did an experienced crew flying an excellent and serviceable aircraft in a controlled manner crash into the side of Mt Erebus in clear visual conditions? Gordon Vette was so convinced that this was not another case of pilot error (as concluded by the Chippindale Report) that he set about finding scientific evidence to prove it. There were two main factors.

First, the aircraft's navigational equipment, unbeknownst to the flight crew, had been reprogrammed – erroneously as it turned out – just before the flight. The effect of this was to put the McMurdo Sound 'waypoint' behind Mt Erebus on the one and only computer-generated flight plan. This meant that the three inertial navigation systems continued to confirm the crew's mental set that they were proceeding, as expected, up McMurdo Sound.

Second, and this was Gordon Vette's very significant contribution, he was able to demonstrate convincingly that 'sector whiteout' and the effect of a layer of stratus cloud (cutting across Mt Erebus) created the compelling illusion of a long flat snowy expanse cut by an horizon at the far end – just what the crew would expect to see had they actually been flying up McMurdo Sound. Gordon was aided in this by his two sons, both of whom were psychology graduates, and by Professor Ross Day, a world-renowned perceptual expert from Monash University. This work is described in detail in a two-volume book entitled *Impact Erebus*.

The 'Vette theory' was offered in evidence before Justice Peter Mahon's Royal Commission on the Erebus tragedy and withstood all cross-examination and legal and regulatory testing without any amendment. It formed a significant part of Mahon's conclusions. Gordon's challenge to the Chief Inspector of Air Accidents (Ron Chippindale), Air New Zealand's senior management and the prime minister of the time was made at great personal cost. He was forced to retire six years early – he was at the time Air New Zealand's most senior check and training captain. The consequential loss of earnings and superannuation was well in excess of one million New Zealand dollars.

Conclusion

I hope I have conveyed something of the extraordinarily inventive means by which the two principal players achieved their heroic rescues: General Gallieni's saving of Paris and Captain Gordon Vette's finding of Jay Prochnow. Neither acted alone, but it is clear that both individuals were the main architects of their remarkable achievements; they alone orchestrated the necessary actions. And this brings me to a somewhat ageist thought: both men were close to the end of their professional lives: General Gallieni was 65 and had only two more years to live, and Captain Gordon Vette, after putting in more than 30 years on the flight deck, was within sight of retirement – although this was hastened by the events described above.

Do age-related factors have a part to play in achieving heroic recoveries? As someone who is well past his prime, I find this thought appealing. However, I will try to give this notion a more challenging appraisal in the next chapter. Chapter 12 seeks to identify the basic ingredients of what it takes to pull failing systems back from the edge of disaster.

Chapter 12

The Ingredients of Heroic Recovery

Now I have come to the moment of truth – or, at least, to the moment of delivery. Now I must match my words to the promises made in the initial chapter. What does it take to make a heroic recovery?

I could simply repeat the titles of the heroic recovery chapters: training, discipline and leadership; sheer unadulterated professionalism; luck and skill; and inspired improvisation. But these smack of motherhood statements. The problem is that it is relatively easy to say why things went wrong – I have spent a lifetime doing it. But when it comes to explaining why things went right, the usual logical expedients – necessity and sufficiency – don't have the same bite, particularly when chance, contextual and individual factors play such a significant part. But how can we frame the answers to the question set out above in a way that can be transferred to others? This is the challenge I have set myself here and in the rest of this book.

This chapter will be structured under three headings: coping with expected hazards; dealing with unlikely (even very unlikely) but possible hazards; and generic qualities, those that could contribute to successful recoveries in any emergency. The problem is that many of these qualities exist within the heads of specific individuals. And there is also the harsh fact that nobody is at their best all of the time. But there is yet another thing: in any emergency no one quality is enough; they have to combine to produce a good result.

Coping with Expected Hazards

Most dangers fall into the category of being expected – or at least known about in advance. Fire, famine, pestilence, injury, drought, poisons, war, crime, falling, shipwreck, drowning, lightning strikes, floods, volcanic eruptions, ferocious creatures, earthquakes, and so on have been familiar to mankind since before recorded history began. When a threat is expected, it can be guarded against to some extent, although anticipating when and where these dangers will strike is much harder.

Evolution dictates that those of us alive today will have been equipped at birth with an impressive array of survival mechanisms. Other defences will have been acquired during development. Yet other ways of looking after ourselves were instilled into us by our families and the occupations we followed. And beyond individuals, most societies have established the agencies necessary for protecting their citizens: the military, police forces, hospitals, ambulance services, and the like. But all of these safeguards and barriers are likely to be imperfect in some respect – this is the basic premise of the Swiss cheese model discussed in Chapter 5.

Only one of the eleven stories of heroic recovery discussed in Part IV falls unambiguously into the expected hazard category – the retreat of the Light Division from cavalry attacks at Fuentes de Onoro. Though it could be argued that the other two military examples – the US Marine Division's retreat from the Chosin Reservoir and General Gallieni's defence of Paris – fall into the same category. If you are a soldier, the possibilities of surprise and defeat are always on the cards. But I will focus here on the Light Division.

For the Light Division, the movements involved in marching and forming square in the face of enemy horsemen had been drilled into them so often they were almost spinal reflexes. Perhaps the length of their retreat across the battlefield – more than three miles – was unusually long, but the component manoeuvres simply had to be repeated over and over. However, their timing was crucial. If the soldiers of the Light Division behaved like a well-rehearsed *corps de ballet* in a long-running performance, their senior officers were not at all like orchestral conductors, merely repeating interpretations of a familiar score. Although the threat remained more or less the same throughout (aside from occasional pieces of horse-drawn artillery), the circumstances were extremely fluid. The officers not only had to keep their nerve in the face of formidable danger, they also needed a high degree of situational awareness and fine judgement in giving their orders. The orders themselves were highly standardised; but their moment of delivery was not – if given too early or too late, they would be destroyed. In this, the officers were more like footballers in a premier league game than musicians.

So what can be generalised from the Light Division's successful retreat in 1811, the closest to a 'set piece' recovery among the 11 case studies? There would appear to be three main elements: the identification and assessment of an expected hazard (something that in this case would have been done a long time before the event); the development, testing and training of a set of countermeasures designed to neutralise the threat (also established long before it was called upon); and an effective and timely way of deploying these countermeasures, a process relying critically on situational awareness. The latter has three components: perceiving the critical elements in the current situation; understanding the significance of these elements; and making projections as to their future status.

All three of these factors are essential for successful recovery; but, of these, situational awareness is the most important. It is certainly the most difficult to acquire and the most universally needed. Whereas the hazards and their countermeasures are specific to particular places and times (it is unlikely that we would need to form square in the 21st century), good situational awareness is a prerequisite for survival in all potentially hazardous domains. Rather than repeating what has been done so well elsewhere, I will direct readers to the very recent book by my friend Rhona Flin and her colleagues. It is called *Safety at the Sharp End* and includes an entire chapter on situational awareness.[1]

Coping with Unlikely but Possible Hazards

Of the 11 recoveries discussed in Part IV, nine fell into the category of improbable but not altogether impossible emergencies. In a roughly decreasing order of likelihood, they are the rescue of

1 Flin, R., O'Connor, P., and Crichton, M. (2008) *Safety at the Sharp End: A Guide to Non-Technical Skills*. Aldershot: Ashgate Publishing, (Chapter 2).

the *Titanic* survivors; the arterial switch operation; the finding of Jay Prochnow; the Marne taxis; the Jakarta incident; the BAC 1-11 windscreen blow out; the Gimli Glider; saving *Apollo 13;* and United 232.

One of the hardest questions to answer in these nine cases is to what extent were the successful recoveries largely due to the character and skills of the main participants? This is not a question that can meaningfully be asked of the successful retreats described in Chapter 8. It is probable that some comparable unit of Wellington's army or another division of US Marines could have exercised their discipline, training, and leadership so as to extricate themselves from their respective emergencies. Could they have done this and still have inflicted so much damage upon the enemy? That is much less certain, but still possible.

Irreplaceable People

Below are the names of those individuals whose presence, I believe, to have been essential for achieving the (relatively) happy outcomes in the unlikely but remotely possible emergencies set out above. I am sure that this list will not be to everyone's liking – but, in this instance, provoking disagreements will be more interesting and probably more instructive than offering bold assertions. I will start with those names least likely to provoke dissent:

- Captain Bob Pearson and First Officer Maurice Quintal (Gimli Glider);
- Captain Al Haynes (United 232);
- Senior First Officer Alastair Atcheson (BAC 1-11);
- Gene Kranz, Bob Legler, Ed Smylie, Chuck Dietrich and John Aaron (mission controllers for *Apollo 13*);
- General Gallieni (the saving of Paris);
- Captain Gordon Vette (ANZ DC-10);
- Captain Arthur Rostron of the *Carpathia.*

There can be little or no doubt about Bob Pearson and Maurice Quintal. The odds of having a skilled glider pilot as the captain and someone who had flown out of Gimli as the co-pilot, the two things necessary to save the stricken aircraft, are almost infinitesimally small.

Although the partial saving of United 232 was a team effort, I believe it was Al Haynes's personality and his cockpit management skills that were the key elements in bringing that about. And it was his inspirational use of the one and three engines that prevented the aircraft from turning onto its back and falling out of the sky at a very early stage in the emergency.

Could anyone other than Alastair Atcheson have wrestled the BAC 1-11 down to a safe landing with his captain half way out of the window? It is possible, but not likely. It was a very contextually-determined episode. After the disaster of the windscreen blow out, the right people were doing the right things with the right pilot at the controls.

Why do I not include the three astronauts among the irreplaceables involved in the saving of *Apollo 13*? It is certainly not to belittle their contribution to the rescue; but any trio of similar ex-test pilot astronauts could, I believe, have done as well – so excellent was their selection and training. But the controllers on the ground were indispensable by virtue of their special knowledge and their extraordinary ability to solve novel problems: Gene Kranz for his impeccable leadership skills; Bob Legler who figured out how to power up the lunar module; Ed Smylie who solved the carbon dioxide problem; Chuck Dietrich for the alignment technique that put the spacecraft into the right entry corridor; and John Aaron who worked out the procedure for repowering the command module prior to re-entry.

Surely some other military commander of Paris could have thought to use civilian vehicles to transport the desperately needed reinforcements to the frontline? No, I don't think so. General Gallieni's contribution to the saving of Paris was unique. First, he saw the opportunity for attacking von Kluck's exposed flank when virtually all of the other senior French generals were thinking about further retreat and regrouping at a safe distance from the German Armies. Second, he made the victory of the Marne possible by using the taxis to solve the twofold problems of a choked-up railway system and the absence of military transport. I believe that it was because of his long experience as a colonial administrator that he was able to break out of the conventional military mindset and come up with a novel and effective solution. Using civilian transport may seem obvious with hindsight, but not at the time – remember that motorised road vehicles had only been around for ten years.

It could be argued that any experienced airline captain with a navigation licence would have been capable of finding Jay Prochnow. That may be so; but I am not convinced. Gordon Vette brought a unique set of qualities to the task: tenacity, ingenuity and very considerable intelligence.[2] Finding Jay Prochnow was not a straightforward navigational exercise: it involved a succession of difficulties, each one requiring ingenious adaptation of known principles to ever-changing and increasingly more critical circumstances. In short, Jay Prochnow was a very fortunate man to have had Gordon Vette overhead.

Finally, what of Captain Rostron? There is no question that another experienced liner captain could have picked up the *Titanic* survivors, but would he have done it so expeditiously? I doubt it. Captain Rostron's seamanship and direction of the rescue operation were quite exceptional. After establishing the relative positions of the *Titanic* and *Carpathia,* he issued an extraordinarily detailed and comprehensive string of orders relating to the discovery, embarkation, welfare, identification and accommodation of the survivors. He had never undertaken a rescue at sea before, yet it was as though he was reading from some prepared check list of things to be done; but there was no check list – except in his head. Like Gordon Vette, he was thinking it through from first principles, and, in his case, relying on old-fashioned Lancashire common sense. Then he zig-zagged his ship between a large number of icebergs at three knots above its design top speed. Some people might say that he was a reckless micro-manager; but I believe he did exactly what was necessary under the circumstances. In my opinion, he was irreplaceable.

Decision-making Styles

It is interesting to contrast Captain Rostron's decision-making and management styles with those of Al Haynes on United 232.

---

2 If further evidence was needed, then consider Gordon Vette's huge contribution to the Mahon Inquiry.

Consider Al Haynes's use of the one and three engine throttles to stabilise the aircraft's attitude. By his own account, Captain Haynes performed these life-saving actions instinctively. The time frame was, of course, quite different from the rescue of the *Titanic* survivors. Al Haynes had only seconds to right the aircraft before it flipped over and crashed. Neither the *Carpathia* nor its crew and passengers were in any immediate danger, at least initially, and there was an estimated four hours before they were likely to be in the vicinity of the *Titanic*'s lifeboats.

Rhona Flin and her co-workers[3] have provided an excellent and up-to-date account of decision-making styles and their conclusion – borne out by the present discussion – is that there are different methods of deciding, and each suits a particular kind of situation. There are four principal types of decision making: intuitive (recognition-primed), rule-based (where rules are available from remembered experience or from procedures), analytical (choice through comparison of options) and creative thinking (coming up with something entirely new to solve a novel problem). Selection of a decision-making mode depends crucially on assessing the situation. When the risk is high and there are tight time limits, as in the case of Al Haynes, intuitive or dynamic decision-making is often the only option.

When the immediate risk is low and there is relatively little time pressure, a more analytical mode – weighing up the options – is appropriate, as in the case of Captain Rostron, at least when it came to his selection among the possible destinations to disembark the *Titanic* survivors. In giving his orders in preparation to picking up the survivors, he would also have used rule-based decision-making. This involves more conscious effort than the intuitive mode, requiring the decision-maker to recall from experience an appropriate course of action, or to get guidance from a manual or a situation-specific procedure.

The creative mode comes into play in wholly unfamiliar situations in which a novel plan of action is required. This would have been used by the *Apollo 13* controllers at Houston, and on United 232, once the differential engine technique had been found to work, and when Captain Dennis Fitch had come on to

3 Ibid, Chapter 3.

the flight deck. I will say more about this below when I come to compare the management styles of Rostron and Haynes.

One of the reasons for including a user's guide to the mind as Chapter 2 was to introduce the notion of metacognition, or thinking about thinking. Knowing something about how your mind works is often very helpful when making decisions in high-risk situations. Our heads are richly stocked with knowledge structures that are called to mind automatically by similarity-matching and frequency-gambling in response to situational calling conditions. Sometimes, as discussed in Chapter 3, these unconscious search processes can lead us into error. But it is more likely that what is called to mind in this way is going to be an appropriate response. Under pressure, the mind 'defaults' to producing something that has been frequently used in this particular situation – and what is frequently used is, by definition, often very useful.

Gary Klein, a pioneer of the naturalistic decision-making approach, has encapsulated this notion in the title of his recent book *Intuition at Work: Why Developing Your Gut Instincts Will Make You Better at What You Do.*[4] As Rhona Flin and her colleagues have pointed out, the emphasis in cognitive skills training – at least in the military – is shifting from 'what to think' to 'how to think'.[5]

Management Styles

Captain Haynes described his philosophy of cockpit management very clearly in the following statement:

> The bottom line for pilots is that you have resources available to you. Use them as team members – you are not alone up there. If you do have a co-pilot, listen to her or to him. They are sure to have some advice for you. There were 103 years of flying experience in that cockpit when we faced our end and they came through to help – but not one minute of those 103 years had been spent operating an aircraft in the way we were trying to fly it. If we had not worked together, with everybody coming up with ideas and discussing what we should do next and how we were going to do it, I do not think we would have made it to Sioux City.[6]

---

4 Klein, G. (2003) *Intuition at Work*. New York: Doubleday.

5 Flin, R. *et al*. (2008), pp. 60–61.

6 Haynes, A. *Eyewitness Report: United Flight 232*. (http://www.airdisaster.com/eyewitness/ua232.shtml), p. 4.

It is possible that Captain Rostron never articulated, or even thought about, his management style. He had probably assumed it like a mantle, received from a long line of sea captains. Edwardian Britain had rigid class distinctions, and the master's role was the seaborne expression of those divisions.

On hearing of the *Titanic* catastrophe, he explained the situation to his officers, but did not, as far as we know, call for suggestions as to how to proceed. Had they been offered, he might well have listened and given them consideration, so long as they were in line with his own plan of action. But he was very clear about what had to be done; and he was the boss. He carried the can for the safety of his ship and for the success of the mission. And that is probably how his officers saw it as well.

Successful consultation has two elements: a willingness on the part of subordinates to speak up and a corresponding willingness on the part of the leader to listen. United Airlines had been developing a culture of crew resource management since the early 1980s. It is not very likely that a similar culture of sharing ideas would have been possible on the *Carpathia*, nor, in this case, was it necessary or even perhaps desirable. Modern flight decks and the bridges of Edwardian liners are quite different places with quite different needs – though it is very sad that Captain Smith, the master of the *Titanic*, did not have people to persuade him about the dangers of trying to break a speed record through a forecasted ice field. Perhaps he had – like the White Star Line owners – an unshakeable faith in the unsinkability of his ship.

Just as there is no one best decision-making mode for heroic recovery, so there is no one best management style. It's a question of horses for courses; what is appropriate depends upon the circumstances and the people involved. There is no single formula for effective recovery. But there is a way forward.

Generic Qualities

Although the heroic recoveries were largely brought about by the efforts of exceptional people, there were a number of factors that transcended both the individuals and the specific circumstances. These are discussed below.

Realistic optimism

Realistic optimism is the opposite of despair. It is high on the list of necessary attributes for aspiring heroic recoverers, and it is particularly important when there is a succession of problems, as was the case in so many of these emergencies. What wins out is the stubborn belief that it will be all right in the end.

This cheerful confidence was especially evident among the excellent surgeons discussed in Chapter 9. In a complex arterial switch operation, one surgeon went on and off bypass[7] six times because the baby was unstable during the post-bypass period. Dr Jane Carthey, the human factors specialist who personally observed 165 of these procedures, described the surgeon's behaviour as follows: 'Each time the patient was put back onto the heart lung bypass machine, the surgeon tried a different surgical intervention to resolve one of multiple set of underlying problems. The surgeon seemed to be working on a multiple-cause hypothesis and was flexible in his generation of potential causes of the patient's unstable condition. At no time did he show any outward signs of stress and he maintained a belief throughout that the problem was resolvable'.[8]

Dr Carthey also observed that surgeons who were not confident about their chances of fixing a problem were more likely to develop 'cognitive tunnel vision' and commit fixation errors when they became fixated on an incorrect surgical hypothesis.

Other heroic recoverers were also sustained by a 'never-say-die' view of the world. Al Haynes drew comfort from the fact that his aircraft was over the flat farmlands of Iowa and the weather was clear. If they were not able to reach Sioux City, he was confident that he could put the aircraft down somewhere and that there would be survivors.

Captain Rostron drove the *Carpathia* through waters that were dense with icebergs at three knots over its design top speed. But it was moonlit night with good visibility. He said at the subsequent inquiry in New York: 'I could see the ice. I knew I was perfectly clear.'

---

7 The perfusion machine that pumps blood when the heart is stopped.

8 Carthey, J. *et al* (2003) 'Behavioural markers of surgical excellence.' *Safety Science*, 41: 409–425, (p. 420).

The buoyant attitude of the *Apollo 13* controllers is summed up in the words that Gene Kranz didn't actually say at the time but wished he had: 'Failure is not an option'. Afterwards, he wrote: 'I was pretty much betting that this control team would pull me out of the woods once we had decided to go around the moon.' And Chris Kraft said of the astronauts: 'You wanted people who would not panic under those circumstances. These three guys [Lovell, Swigert and Haise], having been test pilots, were the personification of that theory.'

So sure was General Gallieni that von Kluck had made a bad mistake by exposing his flank to the Army of Paris that he made several frustrating attempts to persuade Joffre and the reluctant British that the Germans could be beaten. Finally, his continuous badgering paid off and the French and British turned and fought. A less confident man would not have persisted.

I am currently reading *Bomber Boys*, a detailed account of the lives and deaths of those who were flight crews in Bomber Command during the Second World War.[9] Their losses in men and aircraft were horrendous. Approximately 50 per cent of their total number were killed; but if one factors in the long period of training (and the crews that did not become operational during the war), the number increases to 60 per cent. Flying on a raid over Germany, especially Berlin, and returning intact (or nearly so) gets very close to being a heroic recovery. It is interesting to see how often terms like 'confidence' and 'self-assurance' crop up in Patrick Bishop's analysis of what makes for good crew members and effective crews. Here is a sample:

- 'But for all Joe's irritating ways, Willie and the rest of the crew felt there was something comforting about him. He had a self-assurance "which made him good to fly with. Looking at him the crew felt they were safe, for anybody who loved himself so wholeheartedly must survive and surely could not come to any harm". Their hunch turned out to be justified. Later on when they were faced with emergencies he was "to prove as capable and resourceful as he was exasperating."'[10]

---

9 Bishop, P. (2007) *Bomber Boys: Fighting Back 1940–1945*. London: HarperPress.

10 Ibid, p. 185.

- 'The crucial element in crew cohesion was confidence. If one member lost the trust of his fellows, everyone's morale withered.'[11]
- 'In the air it was on the shoulders of the skipper that the burden of maintaining morale weighed heaviest. Confidence was the great sustaining quality and Willie Lewis's skipper John Maze had it in abundance.'[12]

Something Old, Something New: Finding the Right Balance

This section explores the balance between what the heroic recoverers brought to the emergency by way of training, knowledge and experience, and what they came up with during the situation itself. What is interesting is the many different ways that the balance between old and new measures are struck to achieve a successful outcome. I will briefly review the 11 cases below:

1. *Light Division (1811).* The successful retreat was achieved by the repeated execution of the well-drilled 'forming square' manoeuvre in the face of an expected hazard – a large force of determined French cavalry. Crucial to this was the fine timing and situational awareness of the divisional officers. The most unusual part of this withdrawal was the distance it traversed. That the Light Division arrived at its destination largely intact was due primarily to long-established training, discipline and the on-the-spot leadership.
2. *1st Marine Division (1950).* The most important ingredients were pre-existing – esprit de corps, an aggressive fighting spirit, an ethic of mutual support, attention to the smallest details, good weapons and excellent air support. The nature of the enemy was half-expected but their fighting style came as a surprise. To cope with this, the divisional leaders adapted their tactics – hold well-defended positions throughout the night and destroy the attacking force at daylight – very effectively. This was a relatively unusual theatre of war for the Marine Corps, more accustomed to intensive fighting on or just beyond beach heads.

11 Ibid, p. 185.
12 Ibid, p. 195.

An organised fighting retreat over 80 or so miles of frozen and mountainous territory was not something that any of these men had encountered before.

3. *Captain Rostron (1912).* The extraordinary professionalism of the master of the *Carpathia* was the dominant feature in the recovery of the *Titanic's* survivors. This, I would guess, derived in large part from his personality and from his long experience as an officer in the Merchant Navy – both of which long pre-dated the rescue. Although he had not experienced an emergency rescue before (just as Captain Al Haynes had not lost an engine in over 30 years of flying), he assembled his list of instructions quickly, comprehensively and with great attention to detail. They were, for most part, applications of common sense and established practices to a known kind of emergency rather than adaptations to novel circumstances. Though squeezing the extra three knots out of his engines and then sailing at this enforced speed through a moonlit ice field were not, I suspect, his usual practice. He took a calculated risk with the safety of his ship in order to reach the survivors as quickly as possible.
4. *Apollo 13 (1970).* Unlike the rescue of the *Titanic's* survivors, almost all the critical phases of recovering the astronauts involved novel adaptations and untried modifications. These, of course, were created on the basis of the controllers' wide experience and intimate knowledge of their specialist areas of the spacecraft, as well as prior simulation exercises. It is clear, however, that the balance was very much in favour of the inspired invention of new measures rather than the application of pre-existing ones.
5. *Jakarta Incident (1982).* Volcanic ash hazards to airliners are not altogether uncommon. Between 1980 and 2004, more than 100 jet aircraft sustained damage after flying through volcanic ash clouds. But only two passenger airline flights have suffered the temporary failure of all four engines, and Captain Eric Moody's BA Flight 09 was the first of these – in his case, though, it was a five-engine failure since one engine failed twice.[13] There were engine restart drills, but the crew were unsuccessful on

13 On 15 December 1989, a KLM Boeing 747 experienced a temporary four-engine flame-out after flying into the ash plume from Mt Redoubt in Alaska. It landed safely at Anchorage although it experienced loss of power when the engines restarted.

a number of occasions. Just as they were heading towards the ocean to ditch, the number four engine started and shortly after the other three came to life. The landing at Jakarta was complicated by the sand-blasted windscreens and lack of visual reference. For the most part, this heroic recovery was achieved by the continued application of restart procedures and superb airmanship, particularly during the final approach and landing. The balance here, therefore, relied primarily upon established skills and procedures, though ditching on water would have been a wholly novel experience. Happily it wasn't needed.

6. *BAC 1-11 (1990)*. Though the initiating event – having the captain sucked halfway through the hole in the windscreen – was entirely novel, the recovery was achieved by a cool head, airmanship, extraordinary multi-tasking and old-fashioned stick-and-rudder piloting skills. So, in the recovery at least, success depended almost wholly upon the old rather than the new.
7. *Excellent surgeons (1995–97)*. In addition to their very considerable technical skills, excellent surgeons achieved their successful compensations of adverse events through a variety of mental attributes: optimism, cognitive flexibility, adaptation, communication style, safety awareness and anticipating problems before they spiralled out of control. Good outcomes depended largely on what they brought to the operating theatre rather than upon novel improvisations.
8. *Gimli Glider (1983)*. This was a classic example of adapting existing skills (side-slipping) and knowledge (the whereabouts of the Gimli airfield) to recovering an exceedingly dangerous emergency. The skills and the knowledge were pre-existing, but their application was entirely novel.
9. *United 232 (1989)*. There was over a 100 years of combined flying experience on the flight deck and this was skilfully exploited by Al Haynes's use of crew resource management techniques. Correcting the aircraft attitude by using the throttles of the existing two engines was an entirely new aircraft handling technique, and the subsequent manoeuverings using their differential thrust was also entirely novel.
10. *General Gallieni (1914)*. His career had been mainly that of a senior colonial administrator with little in the way of recent combat experience. This may have been just the combination

necessary to achieve the 'miracle on the Marne'. As a colonial governor he would have been accustomed to thinking beyond the purely military aspects of a logistic problem. His assessment of the strategic implications of von Kluck's turn was probably aided by his not having the defeatist mindset that comes from having been driven back by a seemingly invincible enemy for the previous two weeks.

11. *Gordon Vette (1978).* The main feature here was the ingenious application of basic navigation principles to a relatively novel situation – at least it was for all those concerned. From the moment of his first HF contact with the Cessna, Gordon Vette set Jay Prochnow a number of simple exercises to establish his relative position to the DC-10 and to Norfolk Island. Getting Jay to turn to face the setting sun, while Gordon did the same, established that the Cessna was south of them; using the finger-counting technique indicated that Jay was west of the DC-10; making VHF contact with the Cessna meant that it was some 200 miles south-west of the DC-10. Then, using 'aural boxing' and the sunset times supplied by Prochnow and Norfolk Island, Gordon and his crew established that the Cessna was some 300 miles south-east of Norfolk Island. Then Jay struck lucky and made contact with oil rig under tow.

Is there a recurring pattern here? Not really. The heroic rescues ranged from being almost exclusively dependent upon well-established techniques (Light Division) to the opposite balance in the recovery of *Apollo 13*, when success was due very largely to novel adaptations conjured up the MCC controllers.

It could, of course, be argued that these two events are not really comparable; but even when we restrict the comparisons to the five aviation cases, there are still marked differences in the old: new ratio. The crews of BA Flight 09 and of the BAC 1-11 employed existing procedures and piloting techniques, whereas the United 232 crew had to invent an entirely novel mode of attitude control, and the Gimli Glider captain adapted existing gliding skills to a wholly new situation.

Neither prior experience nor new techniques seem to be the defining features of heroic recovery; both can be involved, and the appropriate balance depends very largely upon those involved and the nature of the emergency situation. In contrast to errors, or even violations, whose occurrence and forms are relatively predictable, heroic recoveries are much more singular and unforeseeable events. As one might expect, given the enormous variability of the emergencies, there is no one best decision-making or management style, nor is there an optimum balance of the old and the new that could apply to all situations. These preponderances have to be finely tuned to the needs of the prevailing circumstances.

Conclusion

One thing comes through very clearly from this attempt to identify the main ingredients of heroic recovery: if there is one single most important contributing factor it is having the right people in the right place at the right time. Some of these people are potentially interchangeable within organisations – for example, I believe that other BA captains could have recovered the Jakarta incident, and that other astronauts could have brought *Apollo 13* back; and that other units of Wellington's army and the Marine Corps could have made their successful retreats – but for the most part the happy (or relatively happy) outcomes depended upon the unique skills of the people actually on the spot. Most of them were irreplaceable. Many if not most recoveries were achieved as the result of a providential combination of 'grace under fire', acute situational awareness, personality, professionalism, team work and, in certain instances, some unexpected skills. Good fortune played a large role in many rescues. The one certain barrier to effective recovery would be a despairing lack of self-confidence – and this can happen to the best people.

But these individual ingredients did not appear altogether out of the blue. They had to be selected for and then trained, nurtured and supported by the organisations that the heroic recoverers served. We will further examine the interaction between individual and organisational factors in the next chapter.

PART V
Achieving Resilience

Chapter 13

Individual and Collective Mindfulness

Consistency versus Variability

The reduction of unsafe acts has become one of the primary objectives for those who manage and control complex hazardous systems. Errors and violations are viewed, reasonably enough, as deviations from desired or appropriate behaviour. The managers attribute unreliability to unwanted variability. As with technical unreliability, they see the solution as one of ensuring greater consistency of human action and hence of the system performance as a whole. They do this, as we have seen in Part II, by standard operating procedures, automation, and defences-in-depth.

What these technical managers often fail to appreciate, however, is that human variability in the form of timely adjustments, tweakings and adaptations is what preserves imperfect systems in an uncertain and dynamic world. And therein resides one of the many paradoxes of safety management.[1] By striving to constrain human variability to only those behaviours that are judged *a priori* to be both safe and productive, they are also undermining the system's most important safeguards. The heroic recoveries, discussed earlier, testify to this.

A Dynamic Non-Event

The essence of this is captured by Karl Weick's insightful observation[2] that 'reliability is a dynamic non-event'. It is dynamic because processes remain within acceptable limits

---

1 Reason, J. (2000) 'Safety paradoxes and safety culture.' *Journal of Injury Control and Safety Promotion*, 7: 3–14.

2 Weick, K.E. (1987) 'Organizational culture as a source of high reliability.' *California Management Review*, 29: 112–127.

due to moment-to-moment adjustments and compensations by the human operators. It is a non-event because safe outcomes claim little or no attention. The paradox is rooted in the fact that accidents are salient, while 'normalcy' is not.

Weick and his colleagues[3] have challenged the received wisdom that an organisation's reliability depends upon consistency, repeatability and the invariance of its routines and activities. Unvarying performance, they argue, cannot cope with the unexpected. To account for the success of high reliability organisations (HROs) in dealing with unanticipated events, they distinguish two aspects of organisational functioning: cognition and activity.

The cognitive element relates to being alert to the possibility of unpleasant surprises and having the collective mindset necessary to detect, understand and recover them before they bring about bad consequences. Traditional 'efficient' organisations strive for stable activity patterns, yet possess variable cognitions, and these differing perceptions are most obvious before and after a serious accident. In HROs, on the other hand, 'there is variation in activity, but there is stability in the cognitive processes that make sense of this activity'.[4] This cognitive stability depends critically upon an informed culture – or what Weick and his colleagues have termed 'collective mindfulness'.

Collective Mindfulness

Collective mindfulness allows an organisation to cope with unpleasant surprises in an optimal manner. 'Optimal' does not necessarily mean 'on every occasion', but the evidence suggests that the presence of such enduring cognitive processes is a critical component of organisational resilience.

Since catastrophic failures are rare events in well-defended complex systems. Collectively mindful organisations work hard to extract the most value from what little incident and accident data they have. They actively set out to create a reporting culture by commending, even rewarding, people for reporting their

---

3 Weick, K.E., Sutcliffe, K.M., and Obstfeld, D. (1999) 'Organizing for high reliability: processes of collective mindfulness.' In B. Staw and R. Sutton (eds) *Research in Organizational Behaviour*, 21: 23–81.

4 Ibid.

errors and close calls. They work on the assumption that what seems to be an isolated failure or error is likely to come from the confluence of many upstream contributing factors. Instead of localising failures, they generalise them. Instead of applying local repairs, they strive for system reforms. They do not take the past as an infallible guide to the future. Aware that system failures can take a variety of yet-to-be-encountered forms, they are continually on the look out for 'sneak paths' or novel ways in which active failures and latent conditions can combine to defeat or by-pass the defences, barriers and safeguards. In short, collectively mindful organisations are preoccupied with the possibility of failure.

In talks, I often use the grey squirrel as an example of a high-reliability rodent (they abound outside my window). They are probably the smartest creatures on four legs for their size. They have few predators. Dogs and cats are despised. Human beings are largely ignored. But yet they appear to maintain high levels of chronic unease and a twitchy vigilance. They, like the local birds who alternate between pecking the ground and looking around, are a good model for mindfulness.[5]

In this chapter, I want to take the notion of collective mindfulness forward by combining it with individual mindfulness and arguing that both are necessary for maintaining a state of intelligent wariness. My examples in this chapter are drawn mainly from health care – but they are easily generalised to other domains. Let me remind you of some of the reasons why I have selected patient safety as my principal focus here:

- Much of my work over the last ten years has been in the medical area. I mentioned earlier that the patient safety problem is huge and it exists everywhere. About ten per cent of acute care patients are harmed or killed by iatrogenic factors.
- Health carers are not especially error-prone; it is just that their business is extremely error-provoking. The problem is not helped

5 I must exclude free-range chickens that just seem to peck. They are an avian exemplar of unwarranted insouciance. I'm referring to my neighbours' chickens, taken by a vixen, which we mourn on both sides of the dry-stone wall. They were excellent egg layers. Let that be a lesson to all those who claim 'it couldn't happen here'.

by a medical culture that equates error with incompetence. Medical training is based upon the assumption of trained perfectibility. After a long and arduous training, doctors expect (and are expected) to get it right; but they are fallible, just like the rest of us.

- The US Institute of Medicine and the UK's Department of Health have strongly endorsed a systems approach.[6] While this is better than an exclusive reliance on the 'human-as-hazard' model, it has its limitations. The people on the frontline, nurses and junior doctors, have little chance to change the system; but they still need something to help them. This is from where the notion of individual mindfulness arose.

This chapter will be organised around the elements of Figure 13.1. The basic structure is in the 'background'. It shows an organogram of a health-care system beginning with top management and then senior managers, line managers and the clinicians (often junior doctors and nurses) who are in direct contact with patients. The interface is drawn as a straight line, but the reality is that it is very turbulent, full of ups and downs,

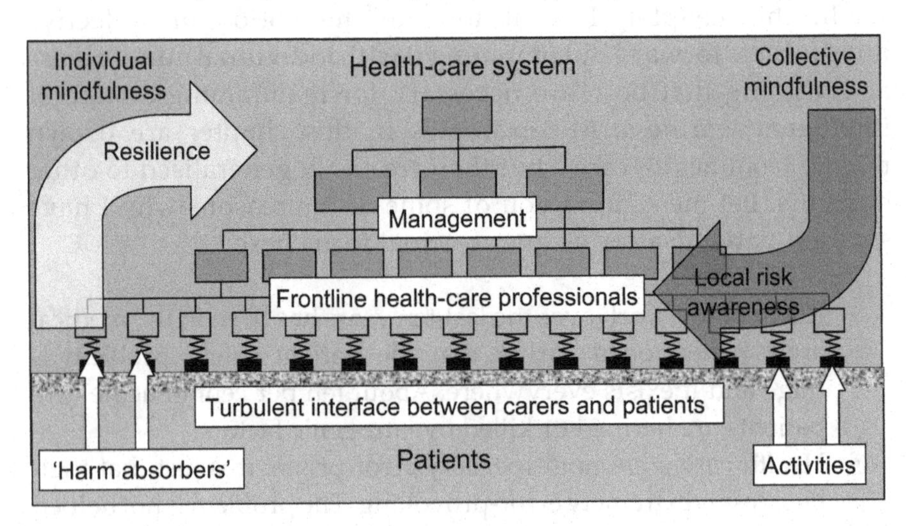

Figure 13.1 Combining individual and collective mindfulness to enhance patient safety

6 Kohn, K., Corrigan, J., and Donaldson, M. (2000) *To Err is Human*. Washington DC: National Academy Press. Donaldson, L. (2000) *An Organisation with a Memory*. London: The Stationary Office.

like a very rough road. The zig-zag lines are analogous to shock-absorbers that soak up the pounding of the bumps. In this metaphor, they are harm-absorbers – they act to minimise threats to patient safety. The rectangles touching the interface are the tasks that sharp-end health carers are required to perform. These are highly variable and frequently involve a diversity of equipment.[7] In what follows I want to focus on the upward arrow to the left of the figure: individual mindfulness leading to systemic resilience.

Individual Mindfulness

In order to explain the development of the notion of individual mindfulness, it would be helpful to begin by referring to a well-documented case study involving the fatal intrathecal injection of the cytotoxic drug vincristine. This has been discussed at length elsewhere,[8] so I will only deal with the events that demonstrate the need for individual mindfulness. It was this case, in particular, that led me to thinking about the ways in which we can enhance 'error wisdom' and risk awareness on the frontline.

An 18-year-old male patient, largely recovered from acute lymphoblastic leukaemia, mistakenly received an intrathecal injection of the cytotoxic drug vincristine. The treatment was given by a senior house officer (SHO) who was supervised by a specialist registrar (SpR). The former was unfamiliar with the usually irreversible neurological damage caused by the intrathecal administration of vincristine, and the latter had only been in post for three days.

It was a requirement that the spinal administration of drugs by SHOs should be supervised by a SpR. This supervisory task fell outside the scope of the SpR's duties at that time, but no one else seemed to be available and he wanted to be helpful. The error was discovered very soon after the treatment and remedial efforts were begun almost immediately, but the patient died just over three weeks later.

---

7 This diversity often exists within equipment items having a common function, as in the case of infusion pumps. There may be as many as 40 different types of infusion devices, having quite different calibrations.

8 Toft, B. (2001) *External Inquiry into the Adverse Incident that Occurred at Queen's Medical Centre, Nottingham, 4th January 2001*. London: Department of Health.

The hospital in question had a wide variety of controls, barriers and safeguards in place to prevent the intrathecal injection of vincristine. But these multiple defences failed in many ways and at many levels. The main 'upstream' defensive breakdowns and absences are listed below – they are also discussed at length elsewhere:[9]

- Administrative and procedural safeguards failed.
- Many indicators, warnings and physical barriers failed.
- There were failures in supervision and instruction of the junior doctors.
- Communication failures and workarounds created defensive gaps.
- There were also collective knowledge failures and false assumptions.

At 17.00 hours, 20 minutes before the drugs were administered, the large majority of the ingredients for the subsequent tragedy were in place. The many gaps and absences in the system's defences had been unwittingly created and were lining up to permit the disaster-in-waiting to occur. Two inadequately prepared junior doctors, each with inflated assumptions about the other's knowledge and experience, were preparing to give the patient his chemotherapy.

It was a Thursday afternoon, normally a quiet time on the ward. The locum consultant was working in his office; the staff grade doctor whom the SpR was supposed to shadow was a part-timer and not on duty that day. The ward sister had gone home. There were no other SpRs available that afternoon. There was no senior medical presence in the vicinity to thwart a sequence of events that was now very close to disaster. To compound the situation further, the patient and his grandmother had arrived unannounced and unscheduled for that particular time. The last 'holes' were about to move into alignment.

The last line of defence was the junior doctors on the spot. The SHO had wanted to administer the drugs in order to gain experience in giving spinal injections. The SpR handed him the

---

9 Reason, J. (2004) 'Beyond the organisational accident: the need for "error wisdom" on the frontline.' *Quality and Safety in Health Care,* 13: ii28–ii33.

syringes. In doing this, he read out the patient's name, the drug and the dose from the syringe label. He did not read out the route of administration. There were also other omissions and errors:

- He failed to check the treatment regimen and the prescription chart with sufficient attention to detect that vincristine was one of the drugs in the packet, and that it should be delivered intravenously on the following day.
- He failed to detect the warning on the syringe.
- He failed to apprehend the significance of the SHO's query – 'vincristine intrathecally?' – on being handed the second syringe.

These errors had grievous consequences. But the SpR's actions were entirely consistent with his interpretation of a situation that had been thrust upon him, and which he had unwisely accepted, and for which he was professionally unprepared. His understanding that he was required to supervise the intrathecal administration of chemotherapy was shaped by the many shortcomings in the system's defences. He might also have reasonably assumed that all of these many and varied safeguards could not have all failed in such a way that he would be handed a package containing both intravenous and intrathecal drugs. Given these false assumptions, it would have seemed superfluous to supply information about the route of administration. It would be like handing someone a full plate of soup and saying 'use a spoon'.

It was clear to see what had lured the SpR into this dreadful situation. But what would have set the alarm bells ringing in his head? There were many indicators: his inexperience, the absence of local supervision, the fact that he was not supposed to engage in clinical work for two weeks and the unscheduled arrival of the patient. The road to disaster was paved with false assumptions – that both drugs in the package were intended for intrathecal delivery, and that the SHO knew the patient – and the lethal convergence of benevolence – the SpR wanted to be helpful, as had the person who put the two drugs in the same package (the local protocol required them to be in separate packages and administered on separate days; but the patient was a poor attender and so unlikely to be persuaded to come to the day ward on two separate occasions).

Acquiring Error Wisdom

Nurses and junior doctors have little opportunity to improve the system defences. But could we not provide them with some basic mental skills that would help them to recognise and, if possible, avoid situations with a high error potential? The 'three-bucket' model shown in Figure 13.2 leads to a possible strategy.

In any given situation, the probability of unsafe acts being committed is a function of the amount of 'brown stuff'[10] in all three buckets. The first relates to the current state of the individual(s) involved, the second reflects the nature of the context, and the third depends upon the error potential of the task. While most professionals will have an understanding of what comprises 'brown stuff' in regard to the self (e.g., lack of knowledge, fatigue, negative life events, inexperience, feeling under the weather and the like) and the context (e.g., distractions, interruptions, shift handovers, harassment, lack of time, unavailability of necessary materials, unserviceable equipment, etc.), they are less likely to know that individual task steps vary widely in their potential to elicit error. For example, omission errors are more likely in steps close to the end of a task, or where there is lack of cueing from the preceding step, or when the primary goal of the task is achieved

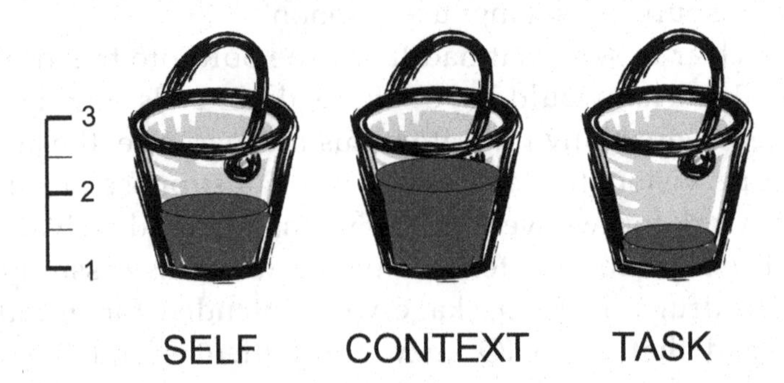

Figure 13.2 The 'three-bucket' model for assessing high-risk situations

10 An internationally understood colour coding – 'brown stuff' is what hits the fan.

before all necessary steps have been completed, and so on. These factors have been discussed at length elsewhere.[11]

Full buckets do not guarantee the occurrence of an unsafe act, nor do nearly empty ones ensure safety (they are never wholly empty). We are dealing with probabilities rather than certainties.

People are very good at making rapid intuitive ordinal ratings of situational aspects. Together with some relatively inexpensive instruction on error-provoking conditions, frontline professionals could acquire the mental skills necessary for making a rough and ready assessment of the error risk in any given situation. Subjective ratings totalling between six and nine (each bucket has a 3-point scale, rising to a total of 9 for the situation as a whole) should set the alarm bells ringing. The buckets are never empty: there is no zero on the scale. Figure 13.3 shows how the ratings might be interpreted by junior staff. Though it must be accepted that in a health-care setting there are other imperatives at work. But more of that later.

There is considerable evidence to show that mental preparedness – over and above the necessary technical skills – plays a major part in the achievement of excellence in both athletics

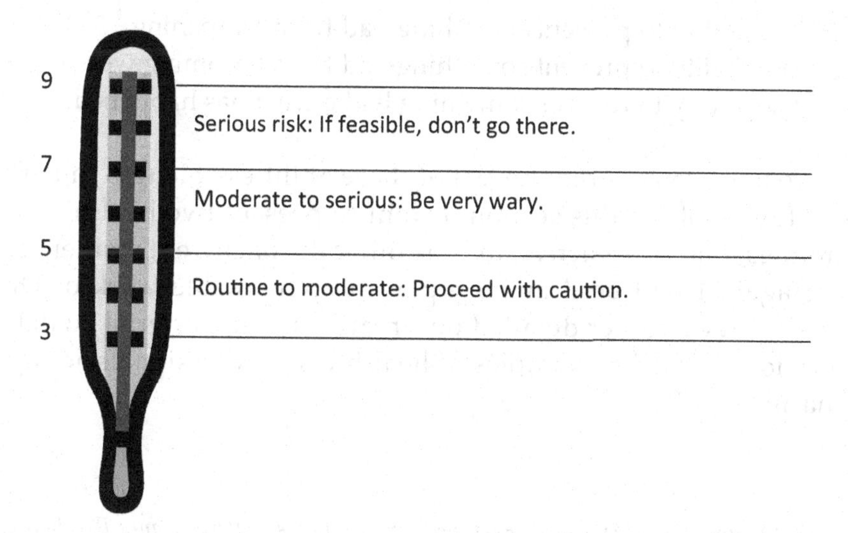

Figure 13.3 How the bucket contents might be interpreted

11 Reason (2004).

and surgery.[12] The 'three-bucket' model and its associated toolkit emphasise the following aspects of preparedness:

- Accept that errors can and will occur.
- Assess the local 'brown stuff' before embarking upon a task.
- Have contingencies ready to deal with anticipated problems.
- Be prepared to seek more qualified assistance.
- Do not let professional courtesy get in the way of checking your colleagues' knowledge and experience, particularly when they are strangers.
- Appreciate that the path to adverse incidents is paved with false assumptions and the lethal convergence of benevolence.

Aspects of Resilience

Ron Westrum, the distinguished American social scientist whose work I mentioned earlier (Chapter 5), pointed out that: 'Protecting the organization from trouble can occur proactively, concurrently, or as a response to something that has already happened.'[13] He argued that each of these is a component of resilience, but they are three distinct entities:

1. The ability to prevent something bad from happening.
2. The ability to prevent something bad from becoming worse.
3. The ability to recover something bad once it has happened.

We have considered the last of these abilities at length in Part IV of this book. In this section, I want to present two health-care examples, one of proactive and the other of concurrent resilience. Once again I will be drawing upon the remarkable work of Dr Jane Carthey and her detailed observations of the arterial switch operation (ASO) for examples of health-care professionals acting as harm absorbers.

---

12 Orlick, T. (1994) *Mental Readiness and its Links to Performance Excellence in Surgery*. Ottawa: University of Ottawa.

13 Westrum, R. (2006) 'Resilience typology.' In E. Hollnagel, D. Woods, and N. Leveson (eds). *Resilience Engineering: Concepts and Precepts*. Aldershot: Ashgate Publishing (p. 59).

Preventing Something Bad from Happening

One per cent of the neonates requiring the ASO procedure are born with a Type B coronary arterial pattern that is extremely difficult to repair. In one case, a pre-operative echocardiogram revealed the presence of the unusual arterial pattern. The surgeon to whom the baby had been referred (Surgeon A) had very little experience of this Type B pattern and had never successfully repaired such a configuration. The surgeon decided to ask another consultant surgeon (Surgeon B), who was known to have had good results with these unusual cases, to perform the ASO procedure on this child. Surgeon B agreed and the operation was performed successfully. Let me quote from Dr Carthey's appraisal: 'He [Surgeon A] may or may not have been successful but the example showed that he used foresight into his own abilities ... to make a decision which optimised the chances of a safe outcome.'[14] This was not an easy decision to make, especially when the prevailing professional culture was that a good consultant surgeon should be able to cope with anything.

Preventing Something Bad from Becoming Worse

During an ASO procedure, the consultant surgeon was distracted and forgot to remove a swab from the pericardial cavity. The scrub nurse repeatedly told the team that her swab count had revealed that one was missing, but her warnings were initially ignored by the surgeon. As he continued the operation, the swab compressed the right coronary artery, though this was not noticed at the time because the baby was on the heart–lung bypass machine. After ten minutes, the scrub nurse forcefully announced that the operation would have to stop because she was no longer going to pass the surgeon suture lines and instruments. The surgeon, unable to continue, had to look for the missing swab. It was successfully recovered and a post-bypass crisis was avoided.

---

14 Carthey, J. *et al.* (2005) *Safety Management Systems, High Reliability Organisations and Resilience Engineering: Implications for Strategies to Improve Patient Safety*. London: National Patient Safety Agency (p. 17).

Foresight Training at the UK National Patient Safety Agency

In 2005, the UK National Patient Safety Agency[15] set up a programme to develop foresight training. It was initially managed by Dr Jane Carthey and was aimed at nurses in the first instance, though it could also be applied to other staff such as patient safety managers, patient safety action teams, risk managers, medical and nursing directors, heads of midwifery and departmental managers. It was organised around the 'three-bucket model' described earlier.

Foresight was defined as 'the ability to identify, respond to, and recover from the initial indications that a patient safety incident could take place'.[16] Foresight training is designed to give nurses (and others) the mental skills necessary to recognise the initial indications that something is amiss. It also provides a chance for staff to share 'war stories' where foresight was used, or could have been used to avert harm to a patient. Furthermore, the process of foresight training also acts as a trigger, raising awareness amongst other clinical colleagues and managers. Nurses have been described as the glue that holds the many disparate parts of the health-care system together; but, being ubiquitous, they are also in a unique position to pass on foresight training messages.

Foresight training sessions are carried out in facilitated groups and use a variety of scenarios, some paper-based and some DVD-based. These scenarios cover acute, primary and mental health settings, and some are potentially relevant for more than one setting. Their purpose is to improve staff knowledge about factors that can make them more likely to be involved in a patient safety incident. Participants assign 'foresight factors' into the self, context and task buckets. The 'three-bucket' framework is intended to help participants think through potential risk factors. The scenarios fall into four categories:

1. *Reflection on action*: these are written scenarios that enable participants to identify and discuss how aspects of self, context

---

15 Established in 2001 as a direct result of the Department of Health publication *An Organisation with a Memory*. (London: DoH, 2000).

16 NPSA (2006) *Foresight Training*. London: National Patient Safety Agency.

and task can contribute to a patient safety incident. They are also asked to consider how prior detection of the 'foresight factors' would have made a difference.

2. *Storytelling*: these are short, written, story-like descriptions of patient safety incidents. They are designed to generate open discussion within the groups about their own experiences of such incidents and how they intervened to prevent harm coming to a patient. Once again, discussions are structured around the 'three-bucket' categories.
3. *Spot the difference*: these use two video (DVD) versions of the same scenario. One describes a situation in which the opportunity for error escalates; the other presents a complimentary scenario in which the harm opportunity decreases. Each pair of videos is designed to provoke a discussion amongst the participants about identifying the 'foresight factors' that contributed to the incident and how they could have been prevented.
4. *Garden path*: these are DVD-based stories that unfold on the screen. The characters in the scenario ask the participants to identify what happens next. Their purpose is to test the participants in their use of foresight and to consolidate the learning from the preceding scenarios.

Feedback from the nurses on the foresight training package was largely very positive. However, a number of potential cautions were expressed, perhaps the most interesting of which was that the notion of foresight training goes against the nursing culture of 'ploughing on to get the job done', and careful thought needs to be given to challenging these attitudes and behaviours. There was also a misperception that the foresight training programme represents a shift away from the system approach to patient safety. I will deal with these issues in the concluding part of the chapter.

Organisational Support

In addition to the upward arrow labelled individual mindfulness, Figure 13.1 has a downward arrow that relates to collective mindfulness. It is clear that programmes designed to improve foresight and 'error wisdom' on the frontline must have strong backing from middle and top managers. It is not enough simply to provide one-off training programmes to instil the necessary mental skills, and then tick the 'done it' box. Mental skills, just like technical skills, need to be continually managed, practised and refreshed. This must be a long-term venture if this, like so many other safety initiatives, is not to wither on the vine.

The organisation must generate a sense of empowerment that allows front line staff to use their judgement and, where necessary, to step back from potentially dangerous situations and to seek help. This is not always possible in a medical setting; but where it is, staff must feel permitted to stop, stand back, think and, where possible, to act to avoid a patient safety incident. Simply urging frontline staff to exercise 'error wisdom' will not work. Both the organisation's culture and its practices must constantly remind them of the hazards and of the need to respect them. Support for individual mindfulness must be embedded in the organisation; without such an infrastructure the programme would simply fade away, getting lost in the press of everyday events.

A good model for organisational support was provided by the Western Mining Corporation (WMC) in Western Australia.[17] They have a programme called 'Take Time, Take Charge' which aims to get workers to stop and think and then take some appropriate action. What makes this happen is that supervisors ask workers each day about situations in which they had taken time and taken charge. These enquiries are prompted by weekly management meetings where the supervisors report these 'take time, take charge' occasions. Those cases deemed to have wider significance are acted upon and the results fed back to the original reporters. Furthermore, WMC has someone at corporate level whose full-time job is to supervise the whole process. Although greater risk awareness of those at the sharp end is the aim, the programme requires the active participation of the managers and supervisors – and is at this level that the success or otherwise of the scheme will be determined.

---

17 Hopkins, A. (2005) *Safety, Culture and Risk: The Organisational Causes of Disaster*. Sydney NSW: CCH Australia Limited (p. 19).

Looking Towards the Future

This section returns to the two dichotomies that have played such a large part in this book. The first relates to the person and system models of safety. The second concerns an often-neglected distinction within the person model: the human as hazard and the human as hero.

Because the empirical foundations of the person model come mainly from event-dependent observations, it is inevitable that human errors and violations are seen as dominating the risks to patient safety. But it is sometimes forgotten that health care would not function at all without the insights, recoveries, adjustments, adaptations, compensations and improvisations performed everyday by health-care professionals.

In their more extreme forms, the person and the system models of patient safety present starkly contrasting views on the origins, nature and management of human error. They have been discussed at length in Part II. A brief reminder: the person model sees errors as arising from (usually) wayward mental processes and focuses its remedial activities upon the erring individual. This view is legally and managerially convenient because it uncouples the responsibility of the error-maker from the organisation at large. The system model, on the other hand, views the frontline fallible person as the inheritor rather than the instigator of an adverse event. Like the patient, people at the 'sharp end' are seen as victims of systemic error-provoking factors and flawed defences that combine, often unforeseeably, to cause unwitting patient harm. The questions that follow from this model are not who went wrong, but which barriers and safeguards failed and how could they be strengthened to prevent a recurrence.

A cyclical progress

In what follows, I will trace a patient safety journey that not only takes account of past and present developments but also anticipates their future consequences. It begins during the 1990s with the widespread acceptance of the human-as-hazard aspect of the person model. It then takes us to the present when a strong endorsement of the system model by many high-level reports has, among other influences, led to an increased awareness of event-causing factors ('resident pathogens') acting at various levels of health-care institutions. But it is also appreciated that systems are slow to change and that we need to provide frontline carers with 'error wisdom' – that is, the mental skills that will help them identify and avoid high-risk situations. It is predicted that the adoption of these error management tools at the 'sharp end' will bring the human-as-hero aspect of the person model into greater prominence. But this can also carry a penalty. Local fixes are likely to lead to the concealment of systemic problems from managers and others with the power to effect more lasting global improvements. It is anticipated that when this process is better understood, there could be a backlash from managers, patients and lawyers that would bring about a reinstatement of the human-as-hazard view, albeit in a more moderate form. At this point, the wheel will have come full circle.

Although these cycles will continue, it is hoped that health-care institutions will learn and mature so that the wide variability evident in the initial go-arounds will gradually diminish to a state when all of these elements can co-exist harmoniously, leading to enhanced resilience and robustness. The main waypoints on this circular path are summarised in Figure 13.4.

The letters A–D in Figure 13.4 identify temporal quadrants in which the transitional drivers operate. Each quadrant is discussed separately below.

Quadrant A: From Human-as-Hazard to Awareness of Systemic Problems

This quadrant covers the period between the late 1990s and the present time. During these seven to eight years, low-level concerns about patient safety have escalated into a widespread acceptance that the problem is huge and that it exists everywhere. In other hazardous domains such a dramatic change usually follows a well-publicised disaster. But, aside from some sentinel events, health care has had no 'big bangs' of this kind. Instead, the wake-up calls came from a flurry of influential reports and epidemiological studies. Perhaps the most influential of these was the Institute of Medicine (IOM) publication released in the latter part of 1999.[18]

---

18 Institute of Medicine (1999) *To Err is Human: Building a Safer Health System*. Washington DC: IOM.

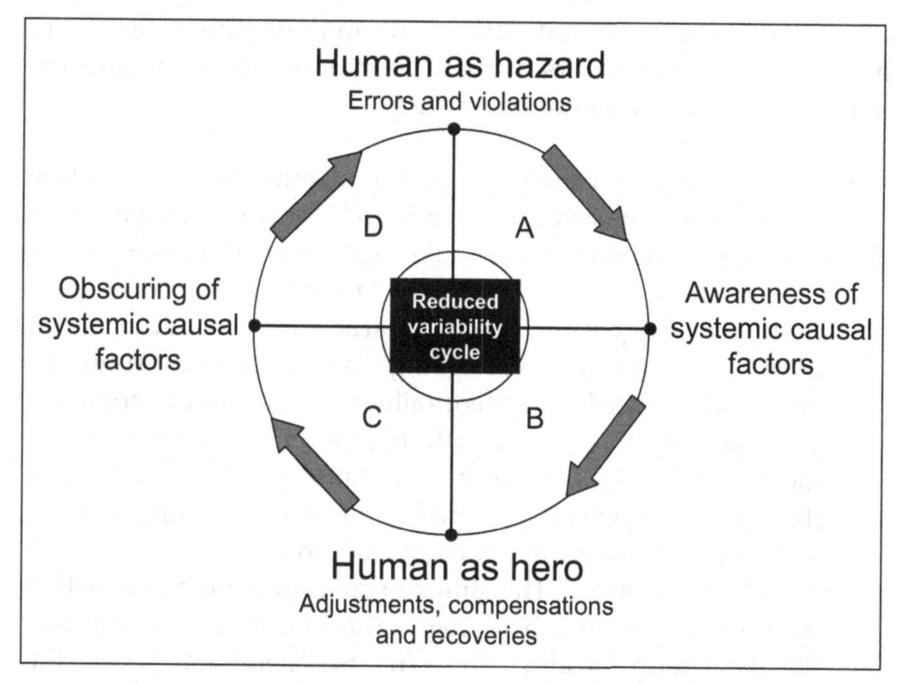

Figure 13.4 Tracing the possible progress of developments in patient safety

The IOM report drew attention the results of the Harvard Medical Practice Study, originally published in 1991,[19] but which had not then received the attention it deserved. Extrapolating from the findings of a patients' notes investigation in New York State, the authors estimated that around 90,000 Americans die each year as the result of medical error – and these numbers have since been revised upwards. Studies in the UK, New Zealand, Australia, Denmark and Canada indicate that around ten per cent of patients admitted to acute care hospitals are killed or harmed iatrogenically.

With the exception of anaesthetists, the human-as-hazard view prevailed among health-care professionals in the 1990s. This naming, blaming and shaming reaction to error was not only intuitively appealing, it was also strongly reinforced by a medical culture that equated fallibility with incompetence or worse. Since then, several factors have combined to create an awareness of the systemic origins of medical error, and a growing recognition that

19 Department of Health (2000).

those practitioners who unwittingly harm their patients are more likely to be the inheritors of institutional 'accidents in waiting' than the sole instigators of bad events:

- As its title suggests, the IOM report condemned the blame culture and strongly advocated a systemic approach to investigating the threats to patient safety. While recognising that human error is perhaps the major contributor to accidents in hazardous systems, the IOM report goes on: '... saying that an accident is due to human error is not the same as assigning blame because most human errors are induced by system failures.'[20] The system approach was similarly promoted by all the high-level international reports mentioned earlier. One of the consequences of such a view is the need to establish standardised incident reporting systems so that organisations might learn from their own mistakes and from those of others. The aim is to create 'organisations with a memory'; that is, ones that recognise not only the proximal causes of patient harm, but also their upstream systemic influences – the 'resident pathogens' or 'error traps' that repeatedly produce the same kinds of errors in different people.
- Following the lead of the anaesthetists (who adopted the system model in the 1980s), health-care professionals now sensibly look to other domains with excellent safety records, particularly commercial aviation, for lessons in risk management. Air accident investigators are required by the International Civil Aviation Organisation (ICAO) to identify which of the many defences, barriers and safeguards failed in order to allow hazards to come into damaging contact with victims, assets or the environment. In each case, the proximal contributions may have involved various kinds of unsafe acts (e.g., errors and violations) but – most significantly – their identification constitutes the beginning rather than the end of the search for causal factors: a search that has, in many cases, led to the discovery of deficiencies in the aviation system as a whole. Moreover, whereas health care is based upon the myth of medical infallibility, aviation, right from its outset a hundred years ago, is predicated on the assumption that people can and will go wrong. Was it Orville or was it Wilbur Wright who devised the first pilot's check list?

20 Institute of Medicine (1999), p. 63.

- Although emotionally satisfying (and driven by strong psychological pressures), as well as being legally and organisationally convenient, many health carers are conscious that a blame culture carries severe penalties. By isolating the erring individual from the context in which the event occurred, it is very difficult to discover the latent conditions that provoked and shaped the unsafe acts. More seriously, a person-oriented approach makes it impossible to identify the recurrent error traps discussed in Chapter 6. Identifying and removing these error traps is a crucial part of risk management. This problem is compounded by the fact that a blame culture and reporting culture cannot easily co-exist – and a functioning incident reporting system is essential for the discovery of error traps. In short, blame has little or no remedial value.

Quadrant B: Restoring the Balance Between the System and the Person Models

Clearly, a systemic approach to the patient safety problem is a vast improvement over the simplistic human-as-hazard approach. But it is not without its drawbacks. Some of the more important of these are listed below:

- Unlike aviation, the activities and the equipment of health care are exceedingly diverse, and while some of the latter may be less sophisticated than in aviation, the interpersonal dynamics are far more complicated, both psychologically and organisationally. Moreover health care has more in common with aircraft maintenance than with the stable routines experienced by commercial jet pilots. Treating patients is a very 'hands-on' activity and, as such, is rich in error opportunities. And although neither the patients nor the clinicians like to acknowledge it, the practice of medicine still has many unknowns and uncertainties. All of these features make the commission of errors more likely, while the fact that patients are already vulnerable people makes the likelihood of causing harm much greater. In addition, the hitherto localised investigation of adverse events makes it harder to learn and disseminate the wider lessons – unlike the extensive publicly reported investigations of aircraft accidents.

- In dangerous industries where the hazards are known and the operations relatively stable and predictable, it is possible to employ an extensive range of automated safety features – or 'defences-in depth'. Whereas some health-care professionals (e.g., anaesthetists, intensivists, radiologists) use comparable automated safeguards, physicians, surgeons and nurses depend heavily on their own skills to keep patients out of harm's way. Patient injury is often just a few millimetres away.
- Nurses, in particular, obtain a great deal of professional satisfaction from fixing system problems at a local level. However, as we shall see below, these workarounds carry a penalty.

Quadrant C: The Downside of Human-as-Hero

At this point in the cycle, we move into the near future and the arguments become more speculative. Nonetheless, there are pointers available from current research to indicate how organisations might become increasingly disenchanted with too much individual autonomy, even of the heroic kind. An example is given below.

Tucker and Edmondson[21] observed the work of 26 nurses at nine hospitals. Their primary interest was in the way that they dealt with local problems that impeded patient care. These problems included missing or broken equipment, missing or incomplete supplies, missing or incorrect information, waiting for a human or equipment resource to appear and multiple demands on their time. On 93 per cent of observed occasions, the solutions were short-term local fixes that enabled them to continue caring for their patients but which did not tackle the underlying organisational shortcomings. Another strategy – used on 42 per cent of occasions – was to seek assistance from another nurse rather than from a more senior person who could do something about the root problem. In both cases, an opportunity for improving the system was lost. In addition, the nurses experienced an increasing sense of frustration and burnout, despite the satisfaction obtained from appearing to cope.

21 Tucker, A.L., and Edmondson, A.C. (2003) 'Why hospitals don't learn from failures: organisational and psychological dynamics that inhibit system change.' *California Management Review*; 45: 55–72.

At a local level, these well-intentioned workarounds appear to smooth out many of the wrinkles of the working day. But from a wider perspective, it can be seen that they carry serious penalties: the concealment of systemic problems from those whose job it is to correct them, and the bypassing or breaching of system safeguards. By their nature, these adverse consequences are not immediately evident. In the short-term, things appear to be working normally. This attitude of 'muddling through' is familiar to all those working in complex systems with less than adequate resources. But, over time, latent pathogens are obscured and others are seeded into the system. This is an insidious process and it is often only after a bad event that we appreciate how these disparate factors can combine to bring about patient harm.

In addition, organisations that rely on – and even encourage – these local fixes come to possess three inter-related organisational pathologies that are symptomatic of poor safety health:

- *Normalisation of deviance*: this is an organisational process whereby certain problems or defects become so commonplace and so apparently inconsequential that their risk significance is gradually downgraded until it is accepted as being a normal part of everyday work. Such a process within NASA was cited as being a factor in both the *Challenger* and *Columbia* shuttle disasters.[22]
- *Doing too much with too little*: this was another factor identified by the Columbia Accident Investigation Board as contributing to the *Columbia* tragedy. It is also a natural consequence of expecting busy frontline health carers to fix local problems as well as giving their patients adequate care.
- *Forgetting to be afraid*: because bad events do not appear to happen very often (at least from the limited perspective of the individual nurse or doctor), health carers can lose sight of the way in which apparently minor defects can combine unexpectedly to cause major tragedies. If there is one defining characteristic of high reliability organisations it is chronic unease, or the continual expectation that things can and will go wrong.

---

22 Vaughan, D. (1996) *The Challenger Launch Decision: Risky Technology, Culture and Deviance at NASA*. Chicago, Il: University of Chicago Press.

Quadrant D: The Reinstatement of the Human-as-Hazard Model

This is likely to occur at some time in the future and so is the most speculative of our transitional periods. Many of the processes described in Quadrant C are likely to be invisible to senior management on a day-to-day basis. It will probably take a number of well-publicised events to bring them to light. But once their significance has been appreciated, it is likely that strong countermeasures will be introduced aimed at limiting the freedom of action of frontline health carers. This backlash is likely to involve a number of top-down measures, the net result of which will be a return to the human-as-hazard model, though it will almost certainly be in a more moderate form.

- There is likely to be a renewed outcry from managers, lawyers and the families of patient victims against 'bad' doctors and nurses. This will receive close media attention and cause predictable reactions from politicians and hospital governing boards.
- Barcoding, computerised physician order systems, electronic health records and automated pharmaceutical dispensing systems have all been implemented to some degree over the past five years. Automatisation takes fallible human beings out of the control loop, at least in the places where errors were commonly made. But this does not necessarily eliminate human error; it merely relocates it. It is probable that part of the backlash against the human initiatives at the sharp end will take the form of more urgent attempts to computerise and automate clinical activities. In the past (and indeed, the present), these innovations have been beset by technical and financial problems; but at this future time, it is likely that many of these difficulties will have been overcome. And the history of complex hazardous systems tells us that one of the ways that management commonly deal with human factors issues is to buy what they see as hi-tech solutions.
- Another favoured countermeasure when dealing with the human factor is to write new procedures, protocols and administrative controls that seek to limit 'sharp end' action to behaviours that are perceived as safe and productive. A fairly safe prediction,

therefore, is that there will be intensified efforts to reduce clinical autonomy. Procedures, protocols and guidelines have an important role to play in safety management, but they are not without their problems as has been discussed at length in Chapter 4.

At first sight, it looks as though the cycle shown in Figure 13.4 has a 'good' sector (right side) and a 'bad' sector (left side). But each is a complex mixture: there is good in the bad and bad in the good. Nothing is wholly black or white; all have potential downsides, all have potential benefits. A better understanding of these issues will permit the anticipation and manipulation of their effects so as to maximise the positives and minimise the negatives.

Reduced Variability

It is expected that as health-care organisations learn more about these processes, variability over the cycle will diminish. The tensions and transitions implicit in the cycle will remain, but their perturbations will become less disruptive. It is hoped that eventually the person and the system models will operate cooperatively rather than competitively. This diminution in variability is represented in Figure 13.5.

It is not possible to step into the same river twice. By the same token, no organisation remains the same. The inner circle in Figure 13.5 represents more moderate perspectives on the issues shown at the outer extremes. It is in this area of reduced variability that we hope to achieve a more mature balance between the system and the person models and, within the latter, between the human-as-hazard and the human-as-hero distinctions. It is further hoped that one of the more enduring products of this equilibrium will be enhanced system resilience. We cannot expect to eliminate human error, technical failures and organisational pathogens altogether (e.g., communication failures, limited resources, economic and governmental pressures), but we can hope to create systems so that they are more resistant to their adverse effects. Greater resilience (unlike zero defects) is an achievable goal.

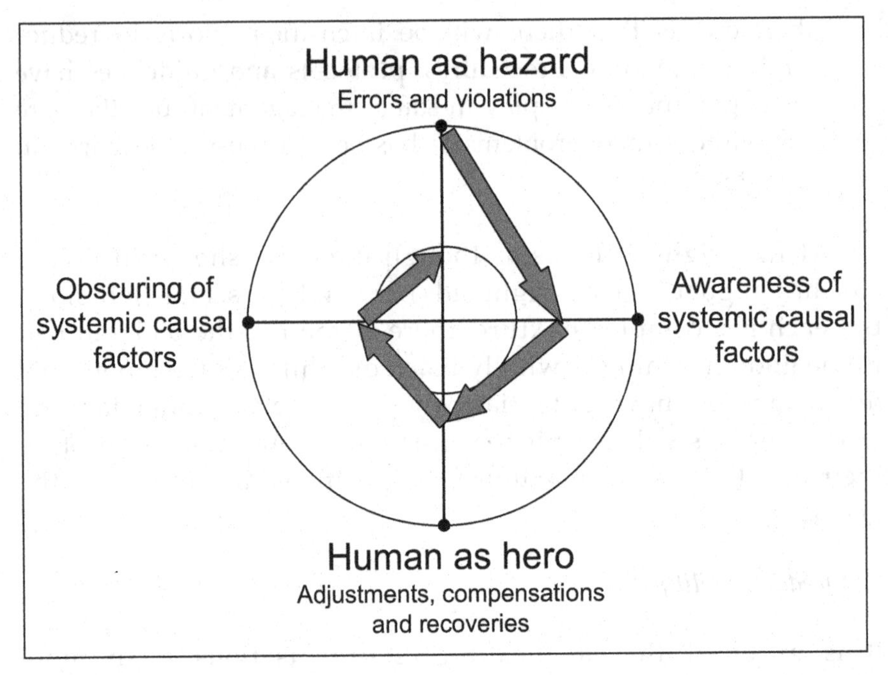

Figure 13.5 Showing the reduction in variability as the cycle progresses

Mindfulness and Resilience

Systemic measures such as standardised equipment, bar-coding, chip-based patient identification, computerised drug ordering and automated dispensing will do much to reduce many of the now commonplace error affordances in patient care, as will establishing the essential components of a safe organisational culture. But while they may be necessary, they are not sufficient in themselves. Ultimately, institutional resilience is an emergent property of the mental and technical skills of those at the sharp end. Richard Cook and his co-authors[23] argued that safe work in the real world depends critically upon 'recognising that hazards are approaching; detecting and managing incipient failure; and, when failure cannot be avoided, working to recover from failure.'

23 Cook, R.I., Render, M.L., and Woods, D.D. (2000) 'Gaps in the continuity of care and progress on patient safety.' *British Medical Journal,* 320: 791–794 (p. 793).

This chapter has sought to reconcile the two dichotomies – person and system, human-as-hazard and human-as-hero – that have concerned us throughout this book. The next and final chapter takes an even broader perspective and presents two metaphors – the safety space model and the rubber-band model – to help us elucidate the rather elusive nature of safety and what is necessary to achieve it at both the organisational and the individual levels.

Chapter 14

In Search of Safety

Introduction

Safety is a term defined more by its absence than its presence. This last chapter seeks to redress the balance by presenting two models that highlight different aspects of the positive and more hidden face of safety. The safety space model deals with goal-setting, intrinsic resistance to operational hazards, the relationship between proactive process measures and reactive negative outcome data, and the importance of both cultural drivers and navigational aids in achieving the maximum practical state of intrinsic resilience. As a corollary to this model, I will show how the three principal cultural drivers – commitment, cognisance and competence – map on to the four 'Ps' of management – principles, policies, procedures and practices – to provide a broad description of what a resilient and safe organisation might look like.

The second model exploits the mechanical properties of a knotted rubber band to enlarge upon the notion of safety as a dynamic non-event. Together the two models provide complimentary views of the nature of safety. The safety space model addresses the more strategic aspects of safety, while the rubber band model focuses on the tactical issues of local control.

What Does the Term 'Safety' Mean?

Like 'health,' the word 'safety' suffers from an imbalance of understanding. Far more is known about its momentary absences than about its longer-lasting presence. We are much better at describing, comprehending and quantifying the occasional deviations from this state, expressed very concretely as accidents, injuries, losses, incidents and close calls than we are at explaining what it means to be safe.

Dictionaries take us no further since they also treat safety as the absence of something. The *Concise Oxford Dictionary*, for example defines safety as 'freedom from danger or risks.' The *Shorter Oxford English Dictionary* gives its meaning as 'exemption from hurt or injury, freedom from dangerousness . . . the quality of being unlikely to cause hurt or injury.'

These everyday uses are of little help to those engaged in the safety sciences or in the management of risk. They neither capture the reality of such activities as aviation, health care and nuclear power generation where the hazards – gravity, terrain, weather, error, hospital-acquired infections, radioactive materials and the like – are ever-present, nor do they tell us much about the nature of the goals that those working within hazardous systems must strive to attain. These people, naturally enough, see their target as the reduction and elimination of harm and losses. But this is only partially within their control. Moreover, in most modern, well-defended technologies such unhappy outcomes are so infrequent as to provide little or no guidance on how to restrict or prevent bad events. Of course, this is not necessarily true for the more 'close encounter' industries, such as mining, transport, health care and construction, but the main focus of this chapter is upon those high-technology activities in which the dangers are potentially great and far-reaching, but where the frequency of adverse events is generally low.

Compared to the natural sciences, where worth is gauged by how much empirical activity a particular theory generates, safety scientists face an additional challenge. As well as provoking interest, a safety-related contribution must also have practical utility. But it can only achieve this if it is readily communicable to those people engaged in the day-to-day business of managing the safety of hazardous operations. Here, as in the behavioural sciences, models, images, metaphors and analogies have an essential part to play. Not only can they convey complex ideas in a concise and digestible fashion (and few enterprises are more complex than the pursuit of safety), they also make it easier for safety specialists, working within dangerous systems, to disseminate these ideas with in their respective organisations.

Such models do not have to be 'true' in the literal sense, nor do they have to be consistent one with another; rather each should tell a story (or draw a picture) that captures some important aspect of an otherwise elusive and mysterious phenomenon. The most useful models also have an internal logic or explanatory 'engine' that highlights the significance of some hitherto unrevealed, or at least unremarked, safety process. The ultimate criterion, though, is a very practical one. Do the ideas communicated by the model lead to measures that enhance a system's resistance to its operational hazards? In short, does it improve safety?

The Two Faces of Safety

Safety has a negative and a positive aspect, though it is mainly the former that claims attention. They are summarised below:

- The negative face is revealed by reactive outcome measures: accidents, fatalities, injuries, loss of assets, environmental damage, patient safety incidents and adverse events of all kinds. Then there are also close calls, near misses and 'free lessons'. All of these are readily quantified and hence much preferred by number-hungry technical managers. These numbers may be convenient and easy to manipulate, but beyond a certain point, they have very dubious validity, as we shall see later.
- The positive face of safety relates to the system's intrinsic resistance to its operational hazards. It is assessed by proactive process measures – indices that reflect an organisation's 'health' both in regard to production and safety. I will say more about these indices later.

The main purpose of the 'safety space' model is to elucidate what exactly is meant by the positive face of safety. But before describing it, let me sneak in yet another metaphor that illustrates the notions of vulnerability and resilience. Figure 14.1 shows a ball-bearing (representing the system) sitting on top of variously shaped metal blocks. Both the ball-bearing and the block are subject to continual jigglings or perturbations that seek to topple the ball-bearing off the block – equivalent to an accident.

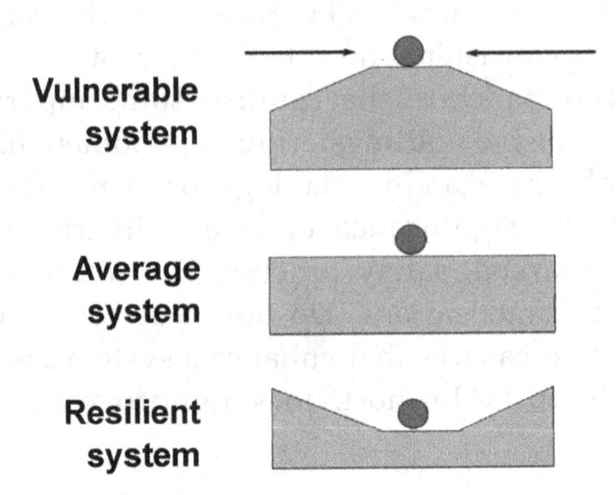

Figure 14.1 Illustrating vulnerability and resistance. The arrows at the top of the figure represent perturbing forces

It is clear that the top ball-and-block are the most vulnerable (easily toppled), while the bottom set are the most resistant. Note, however, that even in this bottom configuration, it is still possible to dislodge the ball. If you prefer a more homely example, think of a tray with an egg and a Pyrex bowl on it. In the vulnerable arrangement, the bowl is inverted and the egg is on top. In the resistant arrangement, the egg is inside the bowl. Perturbations come from tiltings of the tray.

The 'Safety Space' Model

The first model to be described here embodies a navigational metaphor. It presents the notion of a 'safety space' within which comparable organisations can be distributed according to their relative vulnerability or resistance to the dangers that beset their particular activities. They are also free to move to and fro within this space. An important feature of this model is that it seeks to specify an attainable safety goal for real world systems. This is not zero accidents, since safety is not an absolute state; rather it is the achievement and maintenance of the maximum intrinsic resistance to operational hazards.

The model had it's origins in analyses of individual differences in the numbers of accidents experienced by groups of people exposed to comparable hazards over the same time period (we discussed this at some length in Chapter 6). These variations in liability are normally expressed in relation to the predictions of some chance theoretical distribution – the Poisson exponential series. A Poisson distribution looks roughly like the right hand half of a bell-shaped (normal or Gaussian) distribution. But the accident liability distribution is of necessity one-sided; it can only assess degrees of liability. Our concern is with the opposite and previously neglected end of the distribution, most especially with the fact that more than half of the groups assessed in this way have zero accidents. Was this simply due to chance? Were these people simply lucky? It is probable that some of them were. But it is also likely that others possessed characteristics that rendered them less susceptible to accidental harm.

In other words, this unidirectional account of accident liability – discriminating, as it does, degrees of 'unsafety' within a given time period – might actually conceal a bi-directional distribution reflecting variations in personal safety ranging from a high degree of intrinsic resistance to considerable vulnerability. It is a short step from this notional bi-directional distribution of individual accident liability to the 'safety space' shown in Figure 14.2.

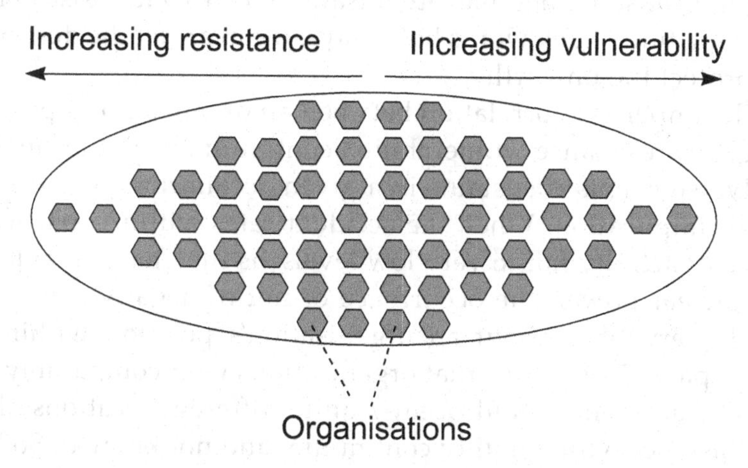

Figure 14.2 Showing a number of hypothetical organisations within the same hazardous domain distributed throughout the safety space

The horizontal axis of the space runs from an extreme of maximum attainable resistance to operational hazards (and still stay in business) on the left to a maximum of survivable vulnerability on the right. Rather than individuals, however, we have plotted the position of a number of hypothetical organisations operating within the same hazardous conditions along this resistance-vulnerability dimension. The space's cigar shape acknowledges that most organisations will occupy an approximately central position with very few located at either extreme.

There will probably be some relationship between an organisation's position along the resistance-vulnerability dimension and the number of bad events it suffers during a given accounting period, but it is likely to be a very tenuous one. If, and only if, the system managers had complete control over all the accident-producing conditions within their organisations would we expect their accident and incident rates to bear a direct relationship to the quality of their efforts. But this is not the case. Chance also plays a large part in accident causation. So long as operational hazards, local variations and human fallibility continue to exist, chance can combine with them in ways that breach the system's defences.[1] Thus, even the most resistant organisations can still have bad accidents. By the same token, even the most vulnerable organisations can evade disaster, at least for a time. Luck works both ways: it can afflict the deserving and protect the unworthy.

The imperfect correlation between an organisation's position along the resistance–vulnerability continuum and the number of adverse events it sustains in a given accounting period has a further implication. When the accident rates within a particular sphere of activity fall to very low levels, as they have in aviation and nuclear power, the occurrence or not of negative outcomes reveals very little about an organisation's position within the safety space. This means that organisations with comparably low levels of accidents could occupy quite different locations along the resistance–vulnerability continuum, and not know it. So how can an organisation establish its own position within the space? In short, what navigational aids are available?

---

1 Reason, J. (1997) *Managing the Risks of Organizational Accidents*. Aldershot: Ashgate Publishing.

Each commercial organisation has two imperatives: to keep its risks as low as possible and still stay in business. It is clear that for any organisation continuing to operate profitably in dangerous conditions, the state of maximum resistance will not confer total immunity from harm. Maximum resistance is only the best that an organisation can reasonably achieve within the limits of its finite resources and current technology. Given these constraints, there are two ways by which it can locate its position within the safety space: from reactive and proactive indices.

Where major accidents are few and far between, the reactive measures will be derived mainly from near miss and incident reporting systems, or 'free lessons.' Such safety information systems have been considered at length elsewhere[2] and will not be discussed further here. We can, however, summarise their likely benefits:

- If the right lessons are learned from these retrospective data, they can act like vaccines to mobilise the organisation's defences against some more serious occurrence in the future. And, like vaccines, they can do this without lasting harm to the system.
- These data can also inform us about which safeguards and barriers remained effective, thus thwarting a more damaging event.
- Near misses, close calls and 'free lessons' provide qualitative insights into how small defensive failures could combine to cause major accidents.
- Such data can also yield the large numbers required for more far-reaching quantitative analyses. The analysis of several domain-related incidents can reveal patterns of cause and effect that are rarely evident in single-case investigations.
- More importantly, the understanding and dissemination of these data serve to slow down the inevitable process of forgetting to be afraid of the (rarely experienced) operational dangers, particularly in systems, such as nuclear power plants, where the operators are physically remote from both the processes they control and their associated hazards.

2 Van der Schraaf, T.W., Lucas, D.A., and Hale, A.R. (1991) *Near Miss Reporting as a Safety Tool*. Oxford: Butterworth-Heinemann.

Proactive measures identify in advance those factors likely to contribute to some future event. Used appropriately, they help to make visible to those who operate and manage the system the latent conditions and 'resident pathogens' that are an inevitable part of any hazardous technology (see Chapter 7). Their great advantage is that they do not have to wait upon an accident or incident; they can be applied now and at any time. Proactive measures involve making regular checks upon the organisation's defences and upon its various essential processes: designing, building, forecasting, scheduling, budgeting, specifying, maintaining, training, selecting, creating procedures, and the like. There is no single comprehensive measure of an organisation's 'safety health'.[3] Just as in medicine, establishing fitness means sampling a subset of a much larger collection of leading indicators, each reflecting the various systemic vital signs

Effective safety management requires the use of both reactive and proactive measures. In combination, they provide essential information about the state of the defences and about the systemic and workplace factors known to contribute to bad outcomes. The main elements of their integrated employment are summarised in Table 14.1.

Table 14.1 Summarising the interactions between reactive and proactive measures

	Type of navigational aid	
	Reactive measures	Proactive measures
Local and organisational conditions	*Analysis of many incidents can reveal recurrent patterns of cause and effect.*	*Identify those conditions most needing correction, leading to steady gains in resistance or 'fitness'.*
Defences barriers and safeguards	*Each event shows a partial or complete trajectory through the defences.*	*Regular checks reveal where holes exist now and where they are most likely to appear next.*

3 Reason (1997).

Navigational aids are necessary but insufficient. Without some internal driving force, organisations would be subject to the 'tides and currents' present within the safety space. These external forces run in opposite directions, getting stronger the nearer an organisation comes to either end.

The closer an organisation approaches the high-vulnerability end of the space, the more likely it is to suffer bad events – though, as mentioned earlier, this is by no means inevitable. Few things alert top management to the perils of their business more than losses or a frightening near miss. Together with regulatory and public pressures, these events provide a powerful impetus for creating enhanced safety measures which, in turn, drive the organisation towards the high-resistance end of the space. However, such improvements are often short-lived. Managers forget to be afraid and start to redirect their limited resources back to serving productive rather than protective ends. Organisations become accustomed to their apparently safer state and allow themselves to drift back into regions of greater vulnerability. Without an 'engine,' organisations will behave like flotsam, subject only to the external forces acting within the space.

Consideration of the 'safety engine' brings us to the cultural core of an organisation. Three factors, in particular, are needed to fuel the 'engine, all of them lying within the province of what Mintzberg called the 'strategic apex' of the system.[4] These driving forces are commitment, competence and cognisance.

Commitment has two components: motivation and resources. The motivational issue hinges on whether an organisation strives to be a domain model for good safety practices, or whether it is content merely to keep one step ahead of regulatory sanctions (see Chapter 5 for a discussion of the differences between 'generative' and 'pathological' organisations). The resource issue is not just a question of money, though that is important. It also concerns the calibre and status of those people assigned to direct the management of system safety. Does such a task put an individual in the career fast lane, or is it a long-term parking area for under-powered or burned out executives?

---

4 Mintzberg, H. (1989) *Mintzberg on Management: Inside Our Strange World of Organizations*. New York: The Free Press.

Commitment by itself is not enough. An organisation must also possess the technical competence necessary to achieve enhanced safety. Have the hazards and safety-critical activities been identified? How many crises have been prepared for? Are crisis plans closely linked to business-recovery plans? Do the defences, barriers and safeguards possess adequate diversity and redundancy? Is the structure of the organisation sufficiently flexible and adaptive? Is the right kind of safety-related information being collected and analysed appropriately? Does this information get disseminated? Does it get acted upon? An effective safety information system is a prerequisite for a resilient system.[5]

Neither commitment nor competence will suffice unless the organisation is adequately cognisant of the dangers that threaten its activities. Cognisant organisations understand the true nature of the struggle for enhanced resilience. For them, a lengthy period without adverse events does not signal 'safe enough'. They see it correctly as a period of heightened danger and so review and strengthen their defences accordingly. In short, cognisant organisations maintain a state of intelligent wariness even in the absence of bad outcomes. This is the very essence of a safe culture.

Figure 14.3 summarises the argument so far. It also identifies the primary goal of safety management: to reach that region of the space associated with the maximally attainable level of intrinsic resistance – and then staying there. Simply moving in the right direction is relatively easy. But sustaining this goal state is very difficult. Maintaining such a position against the strong countervailing currents requires both a skilful use of navigational aids – the reactive and proactive measures – and a powerful cultural 'engine' that continues to exert its driving force regardless of the inclinations of the current leadership team. A good safety culture has to be CEO-proof. CEOs are, by nature, birds of passage: changing jobs frequently is how they got to

---

5 Kjellen, U. (1983) 'An evaluation of safety information systems of six medium-sized and large firms.' *Journal of Occupational Accidents,* 3: 273–288. Smith, M.J., Cohen, H., Cohen, A., and Cleveland, R.J. (1988) 'Characteristics of successful safety programs.' *Journal of Safety Research,* 10: 5–14

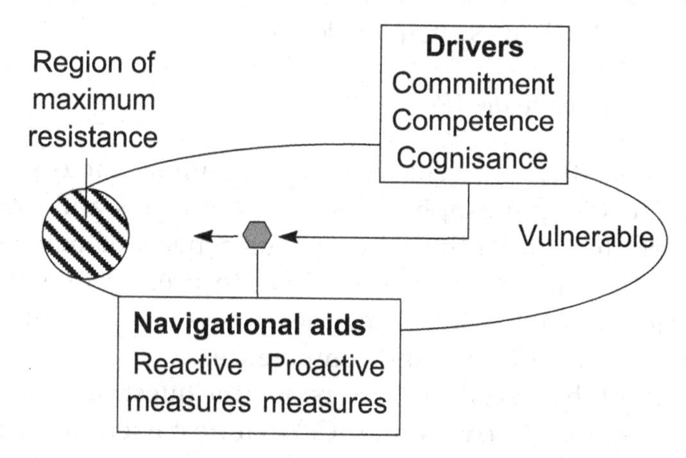

Figure 14.3 Summarising the driving forces and navigational aids necessary to propel an organisation towards the region of maximum resistance

where they are today – and there is no reason to suppose that they are going to behave any differently in the future.

Achieving this practicable safety goal depends very largely upon managing the manageable. Many organisations treat safety management as something akin to a negative production process. They set as targets the achievement of some reduced level of negative outcomes. But unplanned events, by their nature, are not directly controllable. So much of their variance lies outside the organisation's sphere of influence. The safety space model suggests an alternative approach: the long-term fitness programme. Rather than struggling vainly to reduce an already low and perhaps asymptotic level of adverse events, the organisation should regularly assess and improve those basic processes – design, hardware, maintenance, planning, procedures, scheduling, budgeting, communicating – that are known to influence the likelihood of bad events. These are the manageable factors determining a system's intrinsic resistance to its operational hazards. And they, in any case, are the things that managers are hired to manage. In this way, safety management becomes an essential part of the organisation's core business, and not just an add-on.

What Does a Resilient System Look Like?

Mapping the 3Cs on to the 4Ps

Earl Wiener, the eminent American human factors expert, devised the 4Ps (philosophy, policies, procedures and practices) framework[6] to differentiate the various aspects of management activity. I have borrowed the 4Ps here to present one axis of a 3 x 4 table (see Table 14.2). The other axis is made up of the 3Cs (commitment, cognisance and competence.

In each of the 12 cells, we are asking the question: how would each of the cultural drivers (the 3Cs) manifest itself in each of the 4Ps of organisational management? In Cell 1, for example, we are interested in how top-level commitment would reveal itself in the organisation's basic philosophy. In each cell, there are a set of indicators for the influence of the 3Cs upon the three Ps. Collectively, the indicators in the matrix provide a snapshot of what a resilient organisation might look like. The numbers below correspond to the cells in Table 14.2. Some of the contents of the

Table 14.2 Combining the 3Cs and the 4Ps to produce 12 sets of indicators

	Commitment	Cognisance	Competence
Principles (Philosophy	1	2	3
Policies	4	5	6
Procedures	7	8	9
Practices	10	11	12

6 Degani, A. and Wiener, E.L. (1994) 'The four "P"s of flight deck operation.' In N. Johnston, N. McDonald and R. Fuller (eds) *Aviation Psychology in Practice*. Aldershot: Avebury Technical.

cells have a health-care flavour, but these are readily generalised to other hazardous domains:

1. Principles and commitment:
   - Safety is recognised as being everyone's responsibility, not just that of the risk management team.
   - The organisation's mission statement makes safety a primary goal, and this is continually endorsed by the leadership's words, presence, actions and the allocation of resources.
   - Top management accepts errors, setbacks and nasty surprises as inevitable. It repeatedly reminds staff to be wary and vigilant.
   - Safety-related issues are considered at high-level meeting on a regular basis, not just after a bad event
2. Principles and cognisance:
   - Past events are thoroughly reviewed at high-level meetings and the lessons learned are implemented as global reforms rather than local repairs.
   - After some mishap, the primary aim of top management is to identify the failed system defences and improve them, rather than seeking to pin blame on specific individuals at the 'sharp end'.
   - It is understood that effective risk management depends critically upon the collection, analysis and dissemination of relevant safety-related information.
3. Principles and competence:
   - Top management adopts a proactive stance towards safety
     - strives to seek out and remove recurrent error traps,
     - eliminates error-provoking factors in the system,
     - brainstorms new scenarios of failure,
     - conducts regular 'health' checks on organisational 'vital signs'.
   - Top management recognises that error-provoking systemic factors are easier to correct than fleeting psychological states.
4. Policies and commitment:
   - Safety-related information has direct access to the top.
   - Safety management is fast-track not a long-term 'parking lot.'

- Meetings relating to safety are attended by staff from a wide variety of levels and departments.
- Schedulers and planners seek to ensure that teams remain intact when they are known to be effective and where conditions permit.

5. Policies and cognisance:
   - The organisation prioritises clinical goals over non-clinical demands on health-care staff wherever that is possible.
   - Policies are in place to reduce potential sources of non-clinical distraction in clinics, wards and operating theatres.
   - Policies ensure that senior staff are available and present throughout high-risk procedures.
6. Policies and competence:
   - Reporting system policies
     - qualified indemnity against sanctions,
     - confidentiality and/or de-identification,
     - separation of data collection from disciplinary procedures.
   - Disciplinary system policies
     - agreed distinction between acceptable and unacceptable behaviour,
     - peers involved in disciplinary proceedings.
7. Procedures and commitment:
   - The training of junior staff goes beyond the conventional apprenticeship system and procedures are in place to ensure that trainees reach pre-established competency criteria and receive adequate mentoring and supervision.
   - Procedures are in place within the system to facilitate the retraining and continuing professional development of senior staff, particularly with regard to new drugs and techniques.
8. Procedures and cognisance
   - Protocols backed by training in the recognition and recovery of errors.
   - Staff informed by feedback on recurrent error patterns.
   - Shift handovers are proceduralised to ensure adequate communication regarding local conditions.
   - Comparable procedures are in place to ensure safe transitions from the ward or operating theatre to the intensive care unit.

9. Procedures and competence:
   - Clinical supervisors train their charges in the mental as well as technical skills necessary to achieve safe and effective performance.
   - Clinical teams are briefed at the outset of complex or unusual procedures. And, where necessary, they are also debriefed afterwards.
   - The knowledge required to do a job should be shared between procedures, reminders and forcing functions.
10. Practices and commitment:
    - Safety-related issues are discussed by all staff whenever the need arises.
    - Nurses (in particular) should be discouraged from doing 'workarounds' to overcome (often chronic) systemic deficiencies.
    - Rather, they should be rewarded for bringing these problems to the attention of their line management.
11. Practices and cognisance:
    - Frontline personnel (nurses and junior doctors) should be provided with the tools and mental skills necessary to recognise high-risk situations.
    - Junior staff should be empowered to step back from situations for which they have been inadequately trained, where there is no local supervision, and where the conditions are highly error-provoking.
12. Practices and competence:
    - There should be rapid, useful and intelligible feedback on lessons learned and actions needed.
    - Bottom-up information should be listened to and acted upon where necessary.
    - Patient partnering and openness should be encouraged.
    - And, when mishaps occur ...
      - acknowledge responsibility,
      - apologise,
      - convince victims and their relatives that the lessons learned will reduce the chance of a recurrence.

The Knotted Rubber Band Model

The Model Applied to Some Continuous Control Process

What follows is an attempt to elucidate the phrase 'reliability is a dynamic non-event' using the mechanical properties of a rubber band as a model. We are concerned here with the actions of someone on the frontline of the system who has control over some process or piece of equipment.

Imagine a rubber band knotted in the middle. The knot represents the system-to-be-controlled and its spatial position is determined by the horizontal forces exerted on both ends of the band. Three configurations of the knotted rubber band are shown in Figure 14.4.

The stippled area in the centre of the diagram is the safe operating zone. The task of the controller is to keep the knot in this region by countering dangerous perturbations with appropriate compensatory corrections to the other end of the band. The top illustration in Figure 14.4 is a relatively stable state in which moderate and equal tensions on both ends of the band maintain the knot within the safety zone. The middle picture shows an unstable – or unsafe – condition in which an unequal force has been applied to one side of the band, pulling the knot out the safety zone. The bottom configuration depicts a corrected state in which the previous perturbation has been compensated for

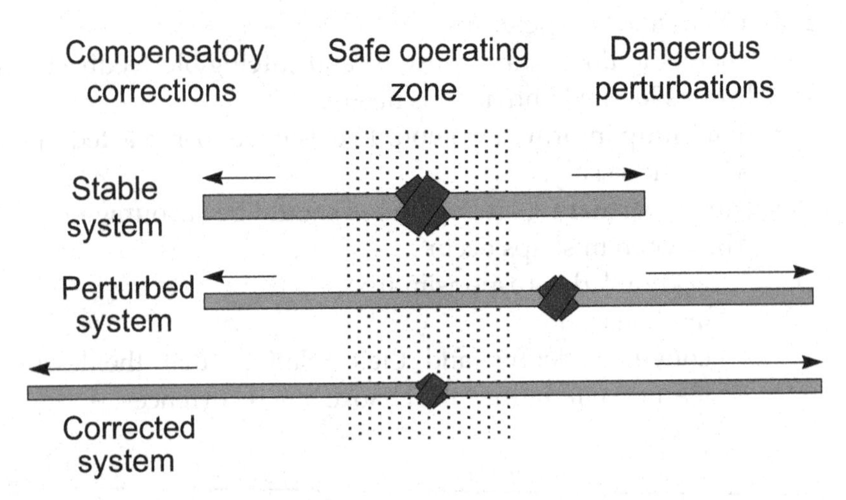

Figure 14.4 Three states of the knotted rubber band

by an equal pull in the opposite direction. There are, of course, many other states, but these are best appreciated by actually manipulating a knotted rubber band yourself.

The rubber band has a further important property. In order to maintain the position of the knot relative to the safety zone, it is necessary to apply an equal, opposite and *simultaneous* correction to any perturbation. Any delay in making this correction will take the knot outside of the safety zone, at least for a short while. I call this the *simultaneity principle.*

In applying this model to complex, highly automated technologies, such as nuclear power plants, chemical process plants and modern commercial aircraft, we should recognise that most of the foreseeable perturbations have already been anticipated by the designers and compensated for by the provision of engineered safety devices. These come into play automatically when the system parameters deviate from acceptable operational limits. This means that the large majority of the residual perturbations – those not anticipated by the designers – are likely to be due either to unexpected variations in local conditions, or to unforeseen actions on the part of the system's human elements – controllers, pilots, maintainers and the like. The latter are likely to include both errors and violations of safe operating procedures (see Chapters 3 and 4).

What are the consequences of the simultaneity principle for the human controllers of complex technologies, taking into account the nature of the residual perturbations just discussed? The first implication is that the timely application of appropriate corrections requires the ability to anticipate their occurrence. This, in turn, demands considerable understanding of what causes these perturbations. That is, it will depend upon the knowledge and experience of the human system controllers regarding, among other things, the roots of their own fallibility. As Weick has argued,[7] these qualities are more likely to be present in systems subject to fairly frequent perturbations (or in which periods of likely perturbation can be anticipated) than in stable systems in which the operating parameters remain constant for long periods of time. Clearly, there will be limits to this generalisation. Just

7 Weick, K.E. (1987) 'Organizational culture as a source of high reliability.' *California Management Review*, 19: 112–127.

as the inverted-U curve (the Yerkes-Dodson law) predicts that optimal human performance will lie between states of low and high arousal, we would similarly expect optimal system performance to lie between the extremes of virtual constancy and unmanageable perturbation.

Support for this view comes from field study observations of nuclear power generation, aircraft carrier flight deck operations and air traffic control.[8] In order to anticipate the conditions likely to provoke error, system operators need to experience them directly, learning from their own and other people's mistakes, as well during simulated training sessions. Error detection and error recovery are acquires skills and must be practised. This need to keep performance skills finely honed has been offered as an explanation for why ship-handlers manoeuvre closer to other vessels than is necessary in the prevailing seaway conditions.[9] Watchkeepers, it was suggested, gain important avoidance skills from such deliberately contrived close encounters.

The Model Applied to the Tension Between Productive and Protective Resources

Figure 14.5 shows the knotted rubber band in a different setting in order to demonstrate its resource implications. Every organisation needs to keep an optimal balance between production and protection (touched upon in Chapter 7 and discussed at length elsewhere).[10] The stippled region is now called the optimal operating zone and on either side there are protective and productive resources, represented as rectangles. The rubber band is a limited resource system. The more it is stretched, the less potential it has for controlling the rubber band – except, of course, by releasing the tension on one or other side.

Three configurations are shown in Figure 14.5. The top one is a balanced state in which the knot is centrally located with considerable potential for corrective action. Configuration A

8 Ibid.

9 Habberley, J.S., Shaddick, C.A., and Taylor, D.H. (1986) *A Behavioural Study of the Collision Avoidance Task in Bridge Watchkeeping*. Southampton: The College of Marine Studies.

10 Reason (1997), Chapter 1.

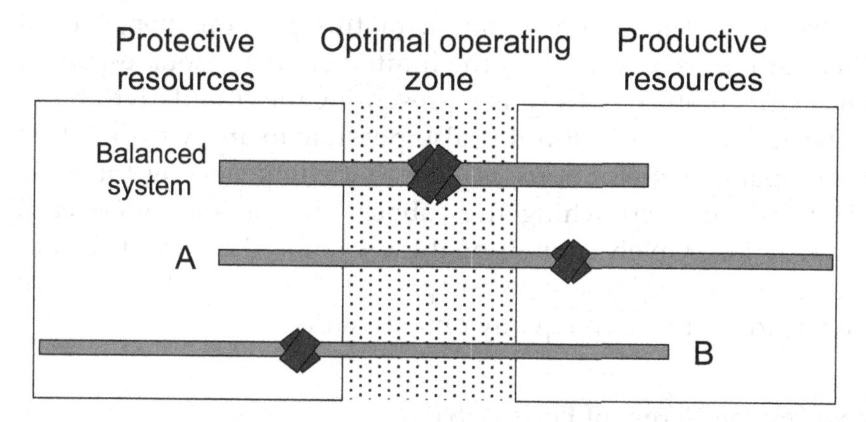

Figure 14.5 Showing the resource implications of the knotted rubber band model

shows an unbalanced state in which the pursuit of productive goals has pulled the knot out of the optimal zone. Configuration B is similarly out of balance, but in the opposite direction. Both configuration A and B have undesirable resource implications. Configuration A provides little or no possibility of compensating for some additional pull in the direction of productive goals, and is potentially dangerous. Configuration B, on the other hand, involves the unnecessary consumption of protective resources and so constitutes a serious economic drain upon the system. The risk in the former case is the unavailability of additional protective resources in the event of an increase in operational hazards; the risk in the latter case is, at the extreme, bankruptcy.

The Model Applied to the Diminution of Coping Abilities

The knotted rubber band has a further application derived from its capacity to become over-stretched and thus lose its potential for correcting the position of the knot. You will recall that in our discussion of the arterial switch operation (Chapter 9), it was noted that the ability of surgeons to compensate for adverse events was inversely related to the total number of events, both major and minor, that were encountered during the procedure. The implication was clear: coping resources are finite. They are used up by repeated stressors.

In the previous consideration of this phenomenon, I used Cheddar cheese to represent the limited coping resources, and a mouse that nibbled it away as representing the cumulative effects of the adverse events. But it is also possible to apply the knotted rubber band model. Let us assume that compensating for these events involves stretching the rubber band to neutralise each perturbation. Given enough of these events, the band becomes over-stretched and is unable to cope with these disturbances until the tension is released equally on both ends.

Defining the Nature of Positive Safety

The purpose of this concluding section is to combine the two models into a single view of safety that does not rely exclusively upon infrequent episodes of 'unsafety'. We will begin by summarising the main features of each model in turn.

Summarising the Properties of the Safety Space Model

- Both people and organisations differ not only in the frequency with which they suffer adverse events, but also in their intrinsic resistance to the hazards of their particular operations.
- It was argued that resistance was a determined rather than a random property. Unlike accidents, that have a large chance component in their causation, the factors contributing to the degree of intrinsic resistance are – to a much greater extent – under the control of those who manage and operate the system. These properties include such generic processes as forecasting, designing, specifying, planning, operating, maintaining, budgeting, communicating, proceduralising, managing, training and the like.
- Because of the chance element, even highly resistant systems can still experience negative outcomes. Safety is never absolute. There is no total freedom from danger. Conversely, even vulnerable systems can escape accidents for lengthy periods. Thus, the relationship between a system's resistance or vulnerability and its accident record, while generally positive over the long run, can be quite tenuous within any specific accounting period.

- Contrary to the spirit of most definitions of safety, it was argued that negative outcome data are imperfect, even misleading, indices of a system's state of intrinsic resistance. This is especially the case when the accident rate is very low or asymptotic – as it is in many contemporary industries.
- It was proposed that an organisation's current level of safety could be represented by its location within a cigar-shaped space, bounded at either end by high degrees of resistance and vulnerability to operational dangers.
- Any organisation is subject to external forces that act to push them away from both end of the space. If they were subject only to these 'tides and currents,' organisations would simply drift to and fro, moving from relative vulnerability to relative resistance, and then back again.
- It was argued that, for any organisation, the only attainable safety goal is not zero accidents, but to strive to reach the zone of maximum practicable resistance and then remain there for as long as possible. For this, each organisation requires both reliable navigational aids and some internal means of propulsion.
- The navigational aids comprise both reactive and proactive data: an effective safety information system that collects, analyses and disseminates information regarding accidents, incidents and near misses that is used in conjunction with regular diagnostic checks upon the system's 'vital signs' and the continuous improvement of those processes most in need of attention at any one time.
- An organisation's engine is essentially cultural. An ideal culture is one that continues to drive an organisation towards the resistant end of the space regardless of the commercial concerns of the current leadership. Three factors are seen to lie at the core of a safe culture: commitment, competence and cognisance (the 3Cs).
- Our consideration of the safety space model concluded with a 12-cell matrix in which indications that each of the 3Cs was influencing each of the 4Ps (principles, policies, procedures and practices) were listed. The matrix as a whole provided a snapshot summary of what a safe and resilient organisation might look like.

Summarising the Main Features of the Knotted Rubber Band Model

- Echoing Weick, the model emphasised the dynamic character of the control actions required to keep a system in a reliable and stable state.
- The mechanical properties of the rubber band model also highlighted the equal, opposite and simultaneous corrections necessary to keep the knot (the system) within the safe operating zone.
- It was argued that most of the foreseeable perturbations (in complex, well-defended systems) will have been anticipated by the designers and corrected for automatically by engineered control devices. Those that remain are likely to arise from local variations and/or from unsafe acts. These human contributions can be both long-standing latent conditions, generated within the upper echelons of the organisation, and active failures (errors and violations) committed by those at the human–system interface.
- In order to achieve the timely correction of these residual perturbations, the system operators must be able to recognise the conditions that foretell their occurrence. To do this, they need to have experienced them (in reality or simulation) and have developed the requisite error detection and correction skills.
- It follows from this assertion that operators of systems subject to relatively frequent disturbances are more likely to possess these skills than those who supervise comparatively stable systems. Systems in which these off-normal conditions are known through direct experience are likely to be safer than those in which this opportunity is largely denied.
- Like the systems it seeks to model, the knotted rubber band is resource-limited. Its corrective potential – beyond a certain point of necessary tension – is inversely related to the degree that it is stretched. When the band is extended near to its breaking point, the only way the knot can be moved is by reducing the tension on one or both sides.
- If it is assumed that the force exerted on one side of the band is primarily protective, while that acting on the other side is essentially productive, two unbalanced system states can be modelled. One is where productive forces hold the knot outside

the optimal operating zone, and the other is the reverse situation in which excessive protective forces have been applied. Both states are potentially dangerous. In the first case, there is the risk of an uncorrected perturbation leading to a bad outcome. In the second, the risk is economic ruin due to an investment in protection that goes beyond that required to counter the operational hazards.

- A third application of the model was in regard to the limited capacity of the coping resources (see the surgeons in Chapter 9). If each perturbation requires a compensatory extension of the rubber band, it eventually becomes over-stretched and so lacks the capacity to counter further disturbances.

How can we integrate these two sets of features into a single coherent account of safety? The task is made easier by the fact that the two models address complementary but somewhat different levels of description. The emphasis of the safety space model is upon the broader strategic aspects of safety, while the rubber band model deals with the more tactical, moment-to-moment, control issues.

The safety space model defines the goal of safety management: the attainment and preservation of a state of maximum practicable resistance to operational hazards. It also indicates, in general terms, to achieve it: that is, the use of reactive and proactive navigational aids and the necessity of a cultural motive force. The dynamics of the knotted rubber band model, on the other hand, are more attuned to the local details of system control, most particularly with the need for anticipating error-provoking conditions, the timing of corrections and a suitable balance between the deployment of protective and productive resources.

Final Words

As I come to write the final words of this book, I am conscious of how ragged and inconclusive it is. I have provided you with little in the way of formulae or prescriptions for safer operation. But, at least, I hope you would be suspicious of anything I (or any consultant) might have offered in that regard. If someone tells you that they have a safe culture you will know to be deeply suspicious, just as you might be if someone told you that they had achieved a state of grace. These are goals that have to be constantly striven for rather than achieved. It's the journey rather than the arrival that matters. Safety is a guerrilla war that you will probably lose (since entropy gets us all in the end), but you can still do the best you can.

I have greatly enjoyed writing about the heroic recoverers. Unfortunately, I have not given you much that could be 'bottled' and passed on to your workforce. Whatever it takes resides largely within very special people. Let's hope you have some such person (or people) when the occasion arises.

Index

Printed in the United States
by Baker & Taylor Publisher Services

Printed in the United States
by Baker & Taylor Publisher Services